This book belongs in the Home Library of
The Owl and the Pussycat
michael Foreman
© Michael Foreman www.myhomelibrary.org
CW00719882

WELCOME TO
SNOOPY'S
WORLD

CREATURES, LARGE AND SMALL

Based on the Characters of Charles M. Schulz

Derrydale Books
New York • Avenel

This 1994 edition is published by Derrydale Books,
distributed by Random House Value Publishing, Inc.,
40 Engelhard Avenue, Avenel, New Jersey 07001

Cover designed by Bill Akunevicz Jr.
Production supervised by Roméo Enriquez

Manufactured in the United States of America

Library of Congress Cataloging-in-Publication Data
Schulz, Charles M.
Creatures large and small / Illustrated by Charles Schulz.
p. cm.—(Snoopy's world)
Originally published as three works: Animals through the ages,
Creatures of land and sea, and Your amazing body.
3rd work originally published: Our incredible universe. 1990.
ISBN 0-517-11902-1
1. Animals—Juvenile literature. 2. Earth sciences—Juvenile literature.
3. Astronomy—Juvenile literature. [1. Space flight. 2. Extinct animals.
3. Body, human—Juvenile literature. [1. Animals. 2. Evolution. 3. Body, Human.]
I. Title. II. Series: Schulz, Charles M. Snoopy's world.
QL49.S27 1994
591—dc20
94-15493
CIP AC

10 9 8 7 6 5 4 3 2 1

INTRODUCTION

In Snoopy's World you'll meet creatures, large and small, weird and wild, long extinct and right in your own home. Have you ever wondered how many bones there are in your body, or why the dinosaurs disappeared, or what an octopus does with all those arms? Charlie Brown, Snoopy, and the rest of the *Peanuts* gang are here to help you find the answers to these questions and many more about *Creatures, Large and Small*. Have fun!

CONTENTS

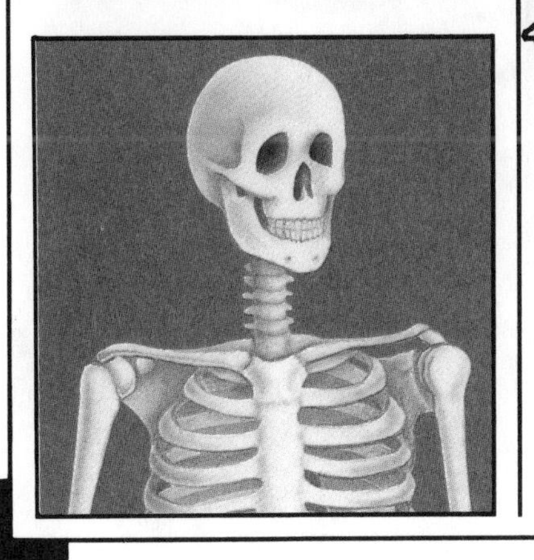

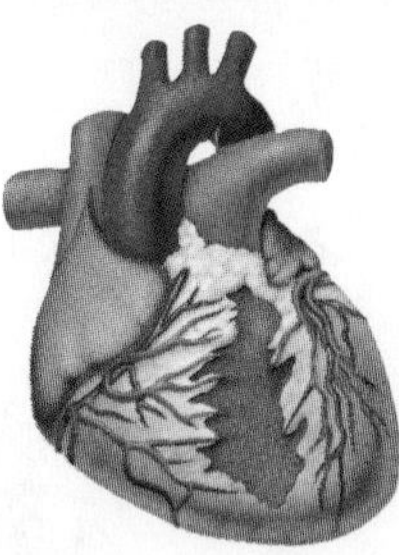

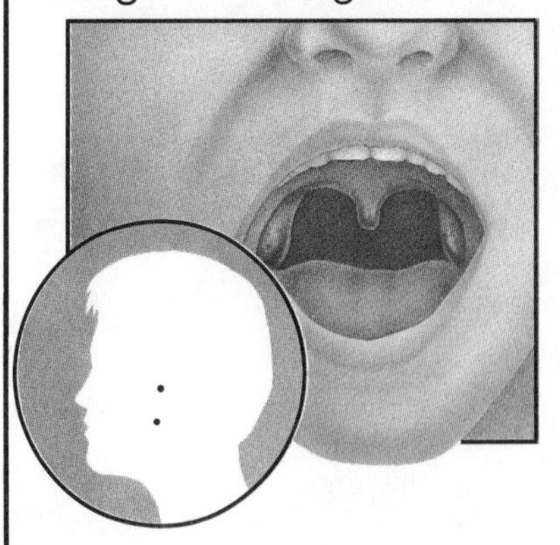

CONTENTS

CONTENTS

There's more to your amazing body than you can see. All parts of you—your bones, your muscles, your skin, your blood, your nerves, your teeth, your hair—are made of cells. These cells are so tiny that you can see them only under a microscope. Your whole body is made up of trillions and trillions of them, and they all come together to make a human being—you!

HOW YOU GROW AND CHANGE

CHAPTER · 1

WHAT YOU'RE MADE OF

What do cells look like?

These are pictures of cells as they look under a microscope.

Not all cells look exactly alike. Different parts of your body are made up of different kinds of cells.

Each kind of cell does a special job that no other kind of cell can do. For example, muscle cells can tighten and relax to make your body move. One kind of blood cell can kill harmful bacteria. Nerve cells send messages to your brain and through all the other parts of your body.

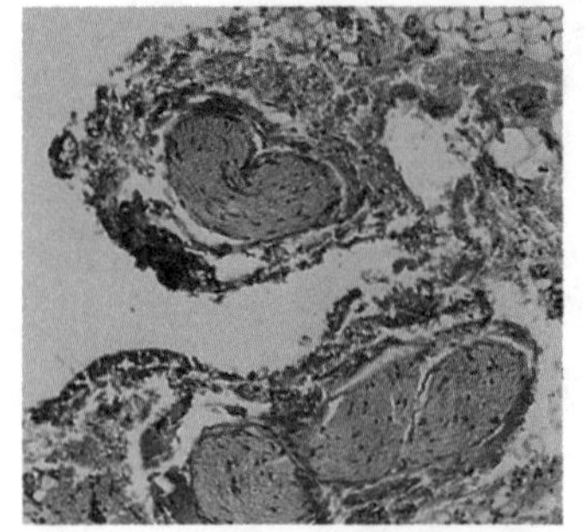

BONE CELLS

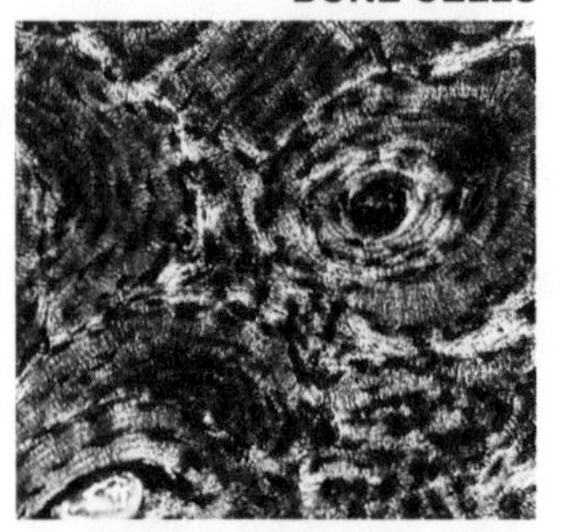

BLOOD CELLS

What makes you grow?

You grow because the cells of your body keep dividing into new cells. When you eat, your cells take in food and grow bigger. Then each cell divides and becomes two cells. Then each of the two cells divides, making four cells, and so on. As the number of cells in your body becomes greater, you grow bigger and bigger.

When did you start growing?

You started growing from just two cells about nine months before you were born. One cell, a sperm cell, came from your father. Another cell, an egg cell, came from your mother. The two cells joined together inside your mother's body. They formed a special new cell called a fertilized egg. This cell was the start of a whole human being—you.

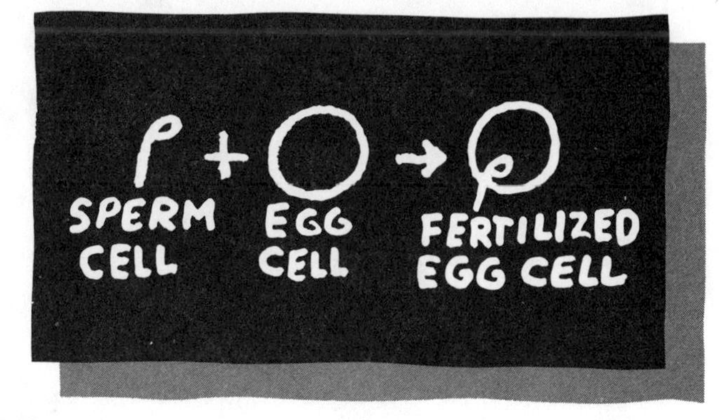

HOW YOU STARTED

How does a fertilized egg cell become a whole person?

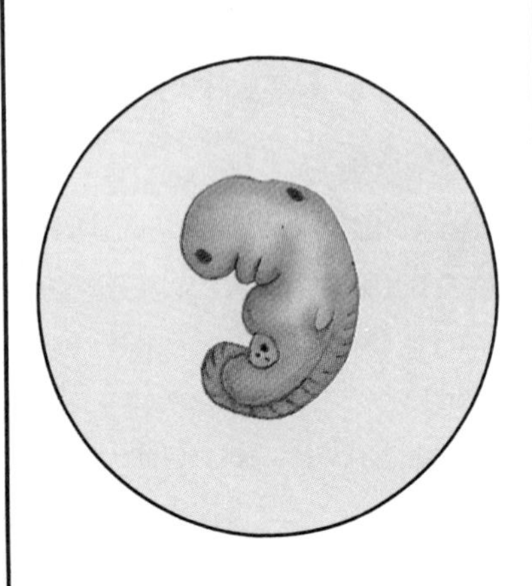

A fertilized egg starts out the size of the head of a pin. It settles inside a special place in the mother's body called the uterus (YOU-ter-us). Then the fertilized egg grows and divides in half. It becomes two cells that are just alike. Then these cells grow and divide. The new cells divide again and again. After a while, not all of them are alike. Some are muscle cells, some are bone cells, some are nerve cells,

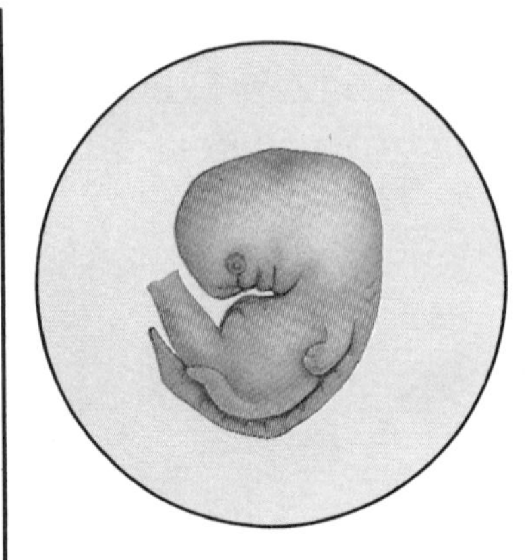

some are blood cells. All the different kinds of cells that make up a human body are there.

About a week after the fertilized egg begins to divide, the new cells start to grow into special body parts—brain,

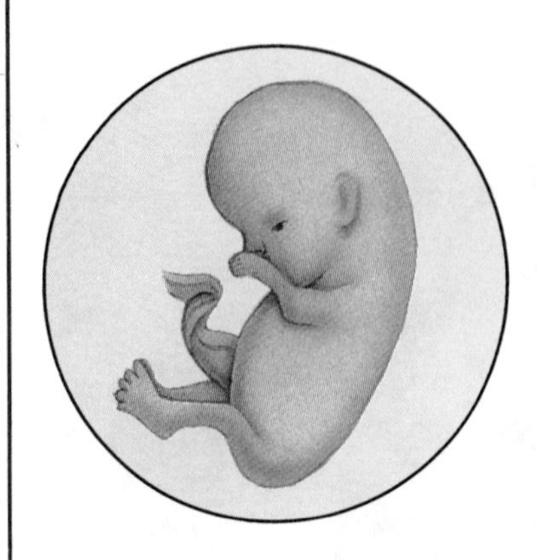

heart, and lungs, for example. After about two months, the developing baby has eyes, ears, a nose, and a mouth. It has tiny legs and arms, too. It has a complete heart that beats and sends blood through its body, but it is still less than an inch long. For seven more months, the baby keeps developing in its mother's body. It

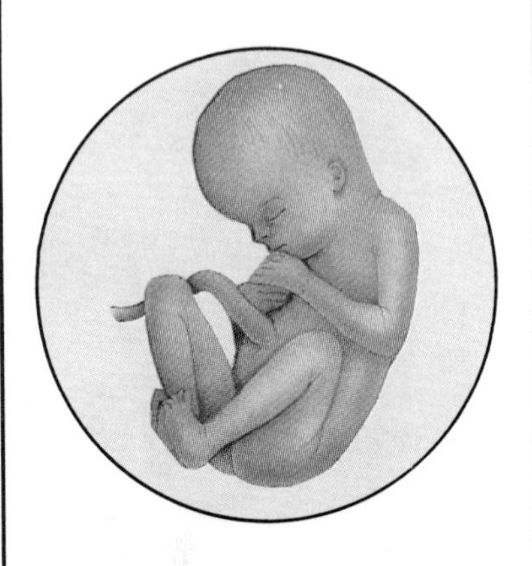

grows bigger and heavier, and looks more and more like a person. At last—about nine months after the fertilized egg began to divide—the baby is born.

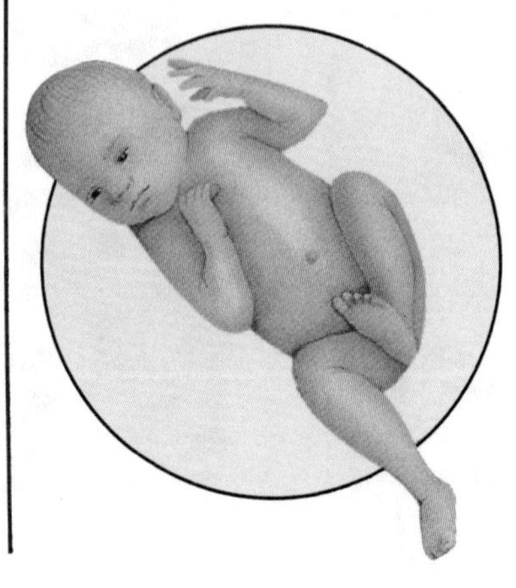

How big is a newborn baby?

A full-term baby, one that is carried the whole nine months, is usually about twenty inches long, and probably weighs between six and nine pounds.

IDENTICAL TWINS

How do twins start growing?

When two sperm cells join with two egg cells at the same time, twins begin to grow. These twins are known as fraternal twins. The two children may not look more alike than any other brothers and sisters, but they are twins.

Twins that start growing from just one fertilized egg cell are called identical twins. After the egg cell is fertilized, it divides completely in half. The two parts are exactly alike. Each goes on to form a separate person. Identical twins look almost exactly alike. Often, only their parents can tell them apart.

Why do you have a "belly button"?

Before you were born, you and your mother were connected in her uterus by a tube called the umbilical cord. Everything you needed to live and grow—including food and oxygen—came to you from your mother through this cord.

After you were born, you no longer needed the umbilical cord because you could eat, drink, and breathe for yourself. So the doctor carefully tied the cord and cut it off. A little dent or bump was left in your belly. This place is called your navel, or belly button.

REACHING YOUR FULL GROWTH

Who grows faster?

Babies break the speed records for growth! When you were a baby you grew faster than you're growing now.

When will you stop growing?

If you are a girl, you will stop growing when you are about 18 years old. If you are a boy, you may keep growing taller for a few more years. After you reach your full height, you may get fatter, but you won't get any taller.

CHAPTER · 2

YOUR SUPER STRUCTURE

If you could look inside your body to see what holds you up, what would you find? Your skeleton, the framework of your body. It gives you your shape and it also protects important parts of you, such as your heart, lungs, and brain.

YOUR BONES

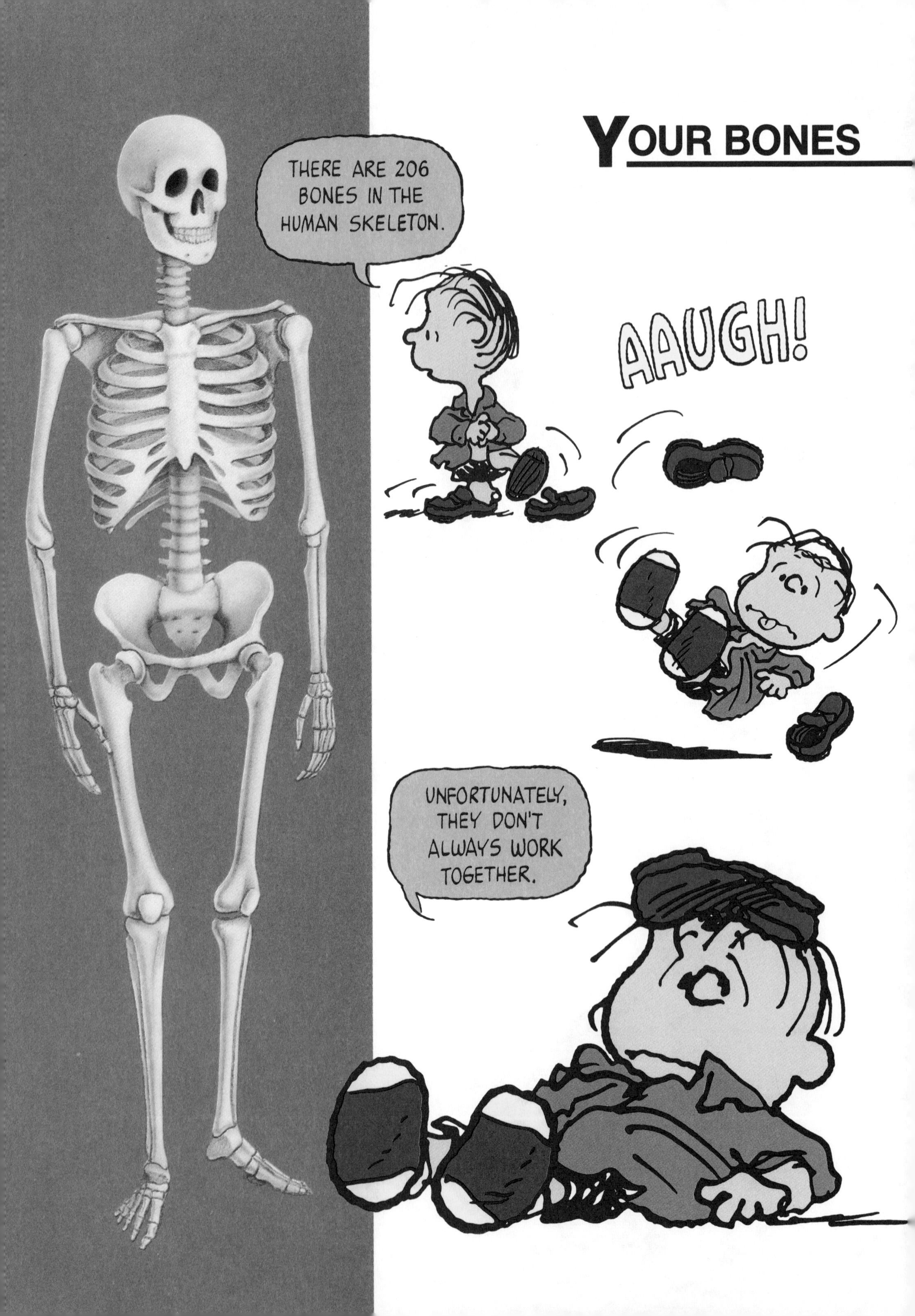
THERE ARE 206 BONES IN THE HUMAN SKELETON.
AAUGH!
UNFORTUNATELY, THEY DON'T ALWAYS WORK TOGETHER.

What do bones look like?

Your bones look somewhat like the beef bone Snoopy is holding, but they are different shapes and sizes. On the outside, they are white and hard and strong. On the inside, they are soft and spongy.

What are the smallest bones in your body?

Three tiny bones in your ear, deep inside your head, are the smallest. The three together are about the size of your thumbnail. These bones look like their names—the hammer, the anvil, and the stirrup.

What are your biggest bones?

Your biggest bones are in your legs. They are your thighbones, or femurs (FEE-mers). If you grow up to be six feet tall, each of your two thighbones will be almost twenty inches long.

Can your bones bend?

No. Your bones cannot bend. You bend your arms, your legs, and other parts of your body at the places where two bones join together. These places are called joints.

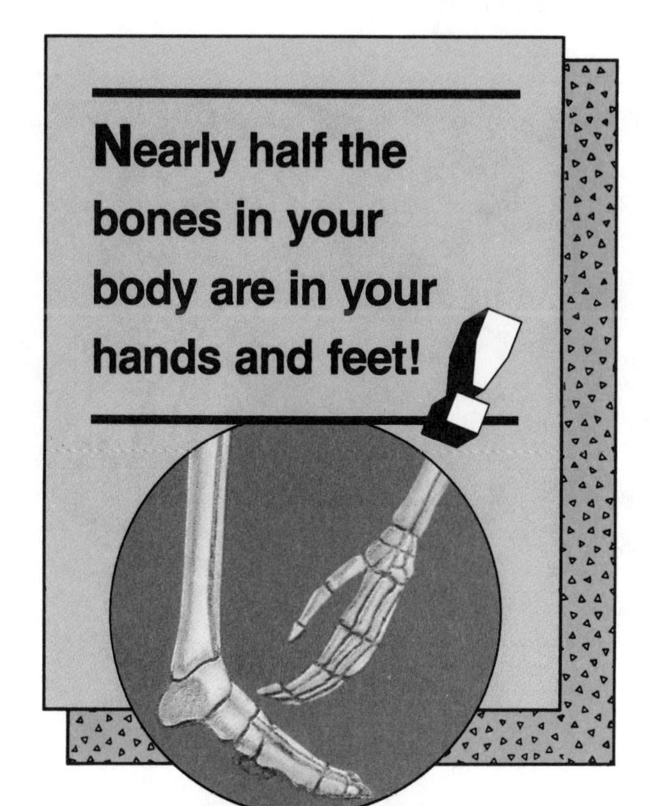

Nearly half the bones in your body are in your hands and feet!

Why do you need a cast when you have a broken arm or leg?

When you say you have a broken arm, you really mean that you have a broken or fractured (cracked) bone. You need a cast to hold the broken parts in place while they heal. A cast keeps the broken ends of the bone from moving around, so that they can grow together again.

YOUR TEETH

Why do your baby teeth fall out?

When you were about six months old, your first set of teeth started to come through your gums. These were your baby teeth. Your baby teeth fall out to make room for larger and stronger teeth.

Everyone grows two sets of teeth.

How does a dentist fix a cavity?

If you have a cavity, the dentist will clean it out to get rid of the bacteria. Then he will fill the hole with silver or porcelain (POUR-suh-lin) so that the cavity cannot grow any deeper and give you a toothache.

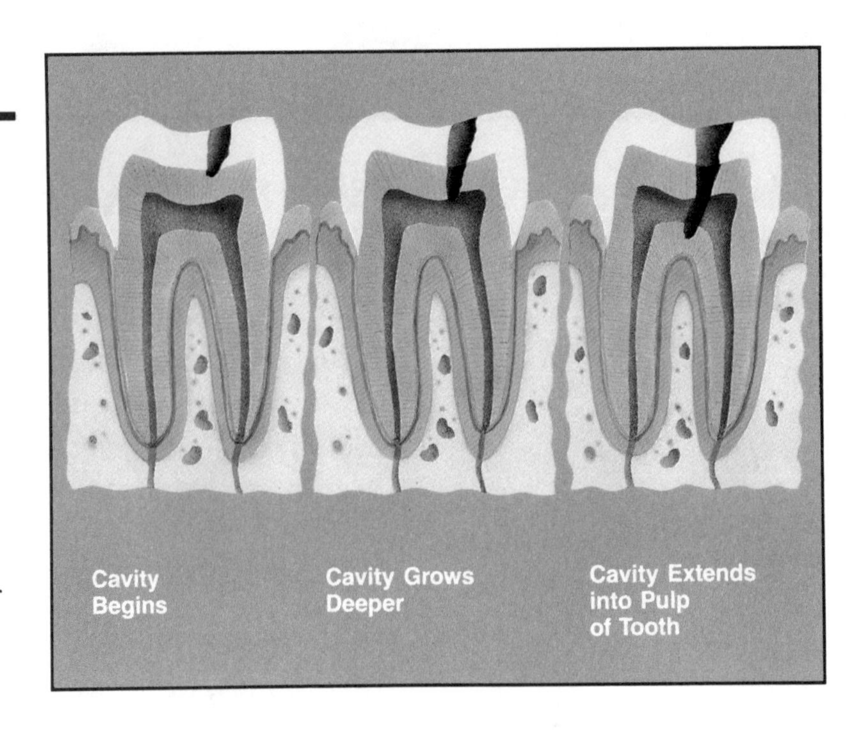

How do you get a cavity in your tooth?

After you eat, tiny bits of food are left between your teeth. If you don't brush away these food bits, bacteria grow on them and start to eat away at the hard outside part of your teeth. The hole that the bacteria make is called a cavity.

A few people lose their second set of teeth and grow a third set!

How can you prevent cavities?

You can help prevent cavities by brushing your teeth after you eat. Always use a toothpaste with fluoride, a chemical that keeps away tooth decay. You should also use dental floss every night to get out all the bits of food stuck between your teeth. Dental floss also removes plaque, a sticky substance on your teeth that can cause disease in your gums.

You can also help prevent cavities by not eating foods with a lot of sugar in them. Cavity-making bacteria grow best in sugar.

You should visit your dentist twice a year. If your water supply doesn't have fluoride, the dentist may apply it directly to your teeth. Thanks to good oral hygiene practices, Americans are getting fewer cavities. Children like you are fighting tooth decay—and winning!

YOUR MUSCLES

Why do you need muscles?

You need muscles in order to move. A muscle is made up of cells arranged in a bundle that can tighten up and get shorter. Then your muscle can relax again and go back to its normal size. When a muscle tightens up, it moves a part of you.

For example, two muscles are at work when you bend your arm at the elbow. These muscles are called the biceps (BY-seps) and triceps (TRY-seps). Each of them is attached to two bones—one at your shoulder and one below your elbow. When you

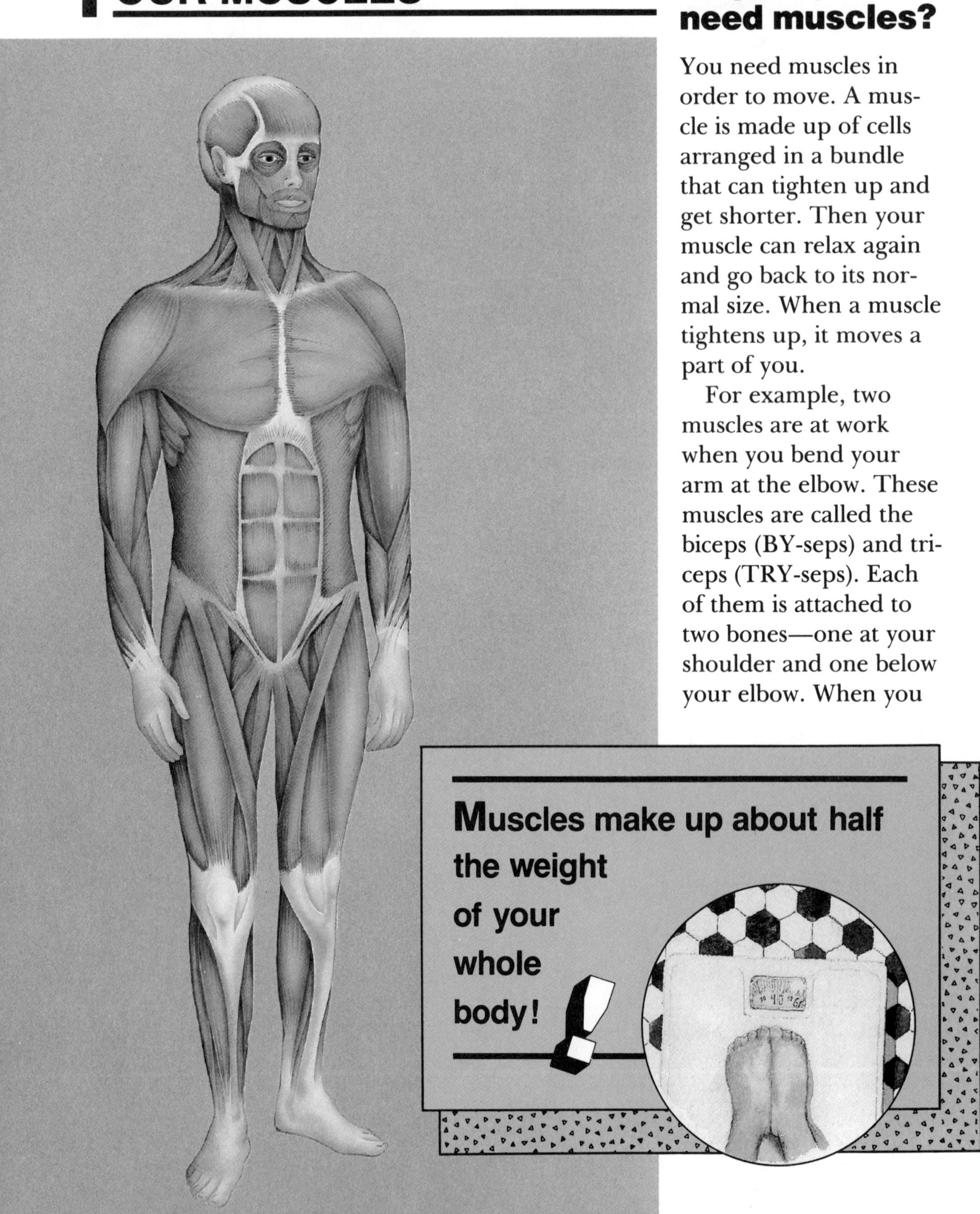

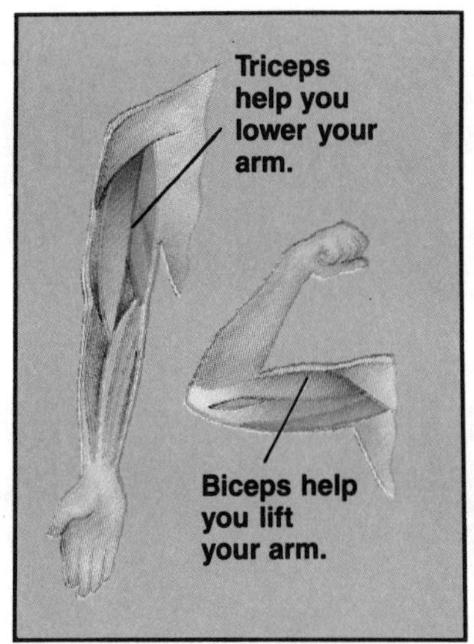

want to bend your arm, say to pick up a pencil on the desk, your brain sends a message to your biceps to tighten up. When the biceps tightens, the muscles get shorter and pull up the lower part of your arm. When you want to straighten out your arm again, your brain sends a message to the triceps to tighten up. This muscle gets shorter and pulls your arm back down. At the same time, your biceps relaxes.

You can easily feel your biceps at work. Put your hand above your elbow. Now bend your arm. You will feel your biceps tightening up.

Why do some people have bigger muscles than other people?

The size of your muscles depends on how much you use them, as well as how big you are. When you do easy things such as sitting, standing, walking, or eating, you use only a small part of each muscle. But when you run, dance, play ball, or swim, you make your muscles work very hard. If you make a muscle work hard very often, it becomes much bigger and stronger. That's why ice skaters have large leg muscles and boxers have large arm muscles.

YOUR SKIN

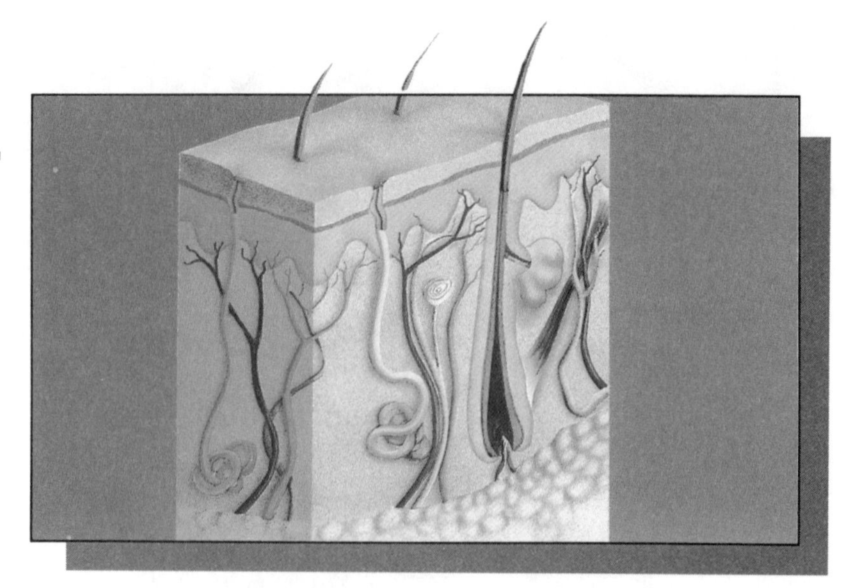

CROSS SECTION OF SKIN

Why do you need skin?

How strange you would look with your insides showing! Your skin covers your body, but it does more than that. It keeps many bacteria out of your body, and so stops them from harming you. It also protects the large amount of water in your body. Without skin, your body would dry out and shrivel up like a raisin.

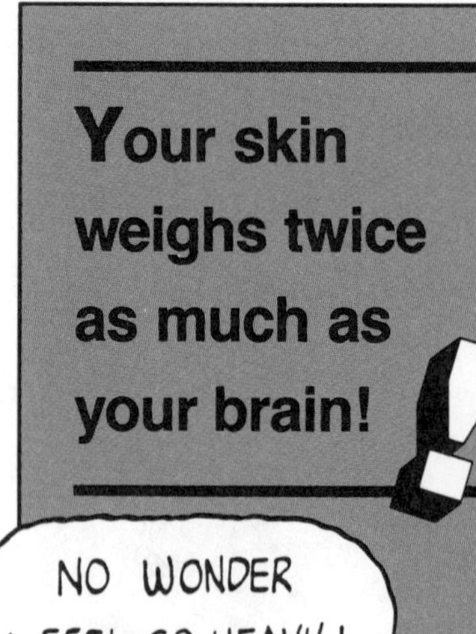

Why does skin get wrinkled?

In the deeper layers of your skin there are tiny muscle fibers that act like a layer of rubber bands. When the skin is stretched, these fibers cause the skin to snap back into place. As a person gets older, some of the fibers disappear, and some lose their snap. Then the skin sags and wrinkles are formed.

Staying out in the sun too much can also cause the skin to wrinkle sooner than it would normally. The sun's rays destroy the elastic fibers in your skin, and that makes it wrinkle. Too much sun can cause serious skin disease, so be sure to protect yourself against sunburn with a sunscreen.

Why do you sweat?

You sweat to cool off. Your body is always making heat. When you exercise, your muscles make extra heat. On a hot day, the sun heats your body, too. If your body did not get rid of some of the extra heat, your temperature would get too high. A very high temperature could kill you. So your body lets heat escape through your skin by sweating.

When you sweat, moisture comes out of your skin. The moisture has heat in it. It evaporates—disappears into the air—carrying the heat with it. Then you feel cooler. When the air is very damp, it cannot take up the moisture and the heat from your skin, so you don't cool off.

Why do some people have freckles?

Freckles are caused by the brown skin pigment called melanin (MEL-uh-nin). All of us have some melanin in our skin. If you have a lot of it, and it is bunched up in spots, you will have freckles, just as Peppermint Patty does.

When sunlight hits your skin, the skin makes extra melanin. So, although you may not have freckles most of the time, you may get them in the sun.

Why do you get "goose pimples"?

"Goose pimples" (or "goose bumps") are tiny bumps that sometimes come out on your skin when you are cold or frightened. If you look closely at the bumps, you will see a hair in the middle of each one. Attached to each hair, inside your skin, is a tiny muscle. When you are scared or chilled, each of these muscles tightens up and gets short. The muscles pull the hairs and make them stand straight up. The skin around each hair is pulled up, too. The result is little bumps. We call these goose pimples because they look like the bumps on the skin of a plucked goose!

Why do people have different-colored skin?

The color of your skin depends on how much pigment, or coloring matter, you have in it. All people have some brown and some yellow pigments in their skins. But different people have different amounts of each pigment. The amount you have depends on the amount your parents have. Because people have such different amounts of the two pigments, many shades of skin color exist in the world. "Black" people have a lot of brown pigment in their skin, and not much yellow. "White" people have a small amount of each pigment in their skin. Asian people have a lot of yellow and a small amount of brown pigment.

What are fingerprints?

Look at the tips of your fingers. Do you see the swirls and loops made by the tiny ridges of the skin? They form the designs that make fingerprints whenever your fingers touch something. Your fingerprints are different from everybody else's in the world. They get bigger as you grow. Otherwise they stay exactly the same all your life.

What is blushing?

The way you feel can affect your body. Sometimes, when you feel embarrassed or ashamed, you blush. Your face and neck look red and feel very warm. Tiny blood vessels in your skin get larger and bring more blood to the top part of your skin. The blood shows through your skin and makes it look red. The blood brings heat with it, so your face and neck also feel warm.

YOUR HAIR AND NAILS

How fast does the hair on your head grow?

In a month, your hair grows about one-quarter of an inch. Even when you stop growing taller, your hair will still keep growing. It grows faster in the summer than in the winter. It grows faster during the day than at night.

Why doesn't it hurt when your hair and nails are cut?

Both hair and nails have no nerves in them. And you cannot feel pain—or anything else—without nerve cells to send messages to your spinal cord and brain. That's why you can cut your hair and nails and never feel a thing.

CHAPTER · 3

Your brain and nervous system are in charge of everything you think about, and just about everything you do. Your brain controls breathing, seeing, hearing, and feeling hungry . . . laughing, reading a book, playing the piano, talking, walking, and crying. Your brain is where the action starts.

YOUR INCREDIBLE BRAIN

YOUR BRAIN AND NERVOUS SYSTEM IN CONTROL

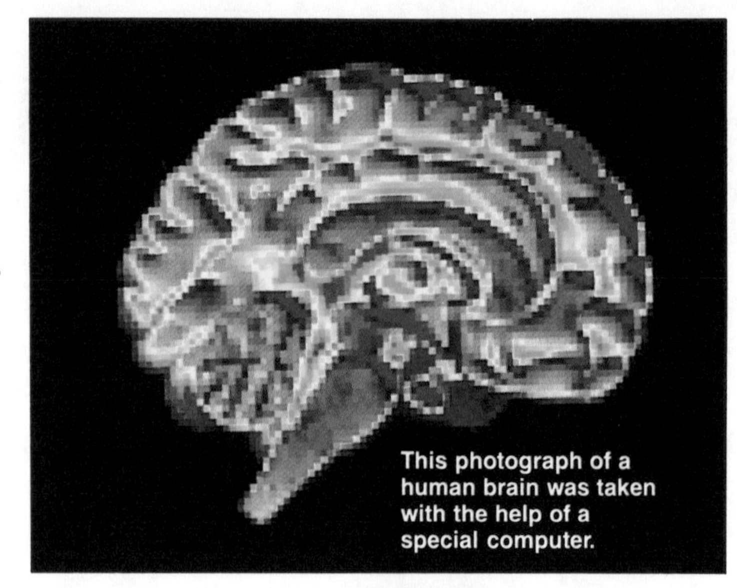

This photograph of a human brain was taken with the help of a special computer.

How does your brain control everything you do?

Your brain is the headquarters of a giant message center called your nervous system. It gets messages from every part of your body by way of special long nerve cells. Your brain then sends its own messages back through the nerves to tell the different parts of your body what to do.

For example, suppose a fly lands on your neck and tickles it. Nerves from your neck send a "tickle message" to your brain. Your brain decides what should be done next. If it decides the tickle should be scratched, your brain sends a message to your arm to lift, and a message to your hand to scratch.

What is your spinal cord?

Your spinal cord is a long cord made of nerve cells. It connects your brain to all the nerves in your body. The spinal cord runs from your brain down your back inside your spine. Most nerve messages pass through your spinal cord on their way to and from your brain.

Some messages travel along your nerves at a speed of 200 miles an hour!

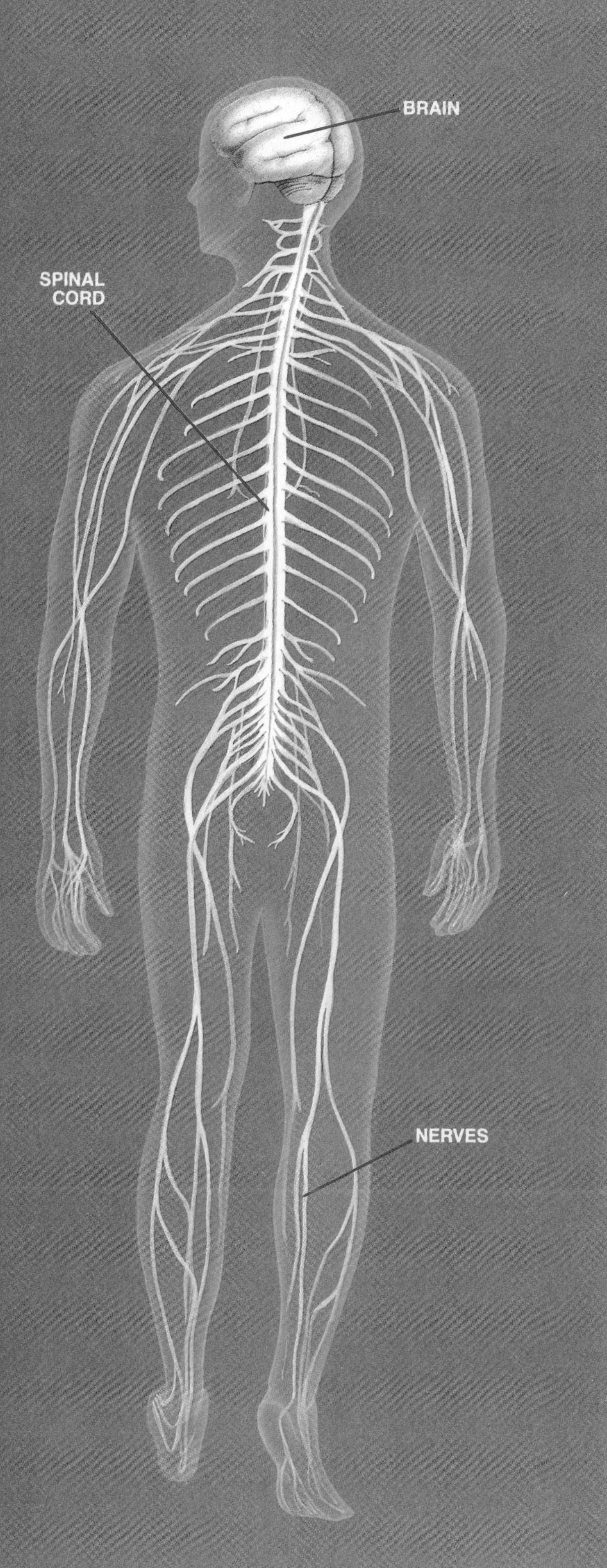

Why must you feel pain?

Have you ever wished that you couldn't feel pain? Well, you are lucky that you *can* feel it. Pain protects you and warns you that something is wrong.

For example, when you have an earache, nerve cells in the ear send a message of pain to your brain. Your brain decides what you should do about the problem—go to the doctor, for instance. If you didn't feel the pain in your ear, you would not know that something was wrong with it. The trouble could get worse and worse, and your ear might end up badly damaged.

Why do you drop a hot potato so fast?

You drop a hot potato even before you can feel the pain that comes from actually burning your hand. As soon as you touch the potato, nerves quickly send a message: "Too hot!" This special danger message goes straight to your spinal cord.

Right away, nerves in your spinal cord answer the message. They don't wait for the message to reach your brain. These nerves make you spread out your fingers so you will drop the potato.

This quick reaction is called a reflex. You don't even have to think about opening your hand. After that, the message goes from your spinal cord to your brain. Your brain makes you realize that the potato was too hot to touch—that touching it caused you pain.

You'll be more careful next time.

What happens when you sleep?

When you sleep, most of your brain and many of your nerves "turn off." Very few messages can be sent to or from your brain. For example, when you are asleep, you don't hear the TV in the next room. But while you sleep, many things continue to go on in your body. Your heart beats, you breathe, and you dream. Your body also replaces worn-out cells.

No one knows for sure why you sleep. Some scientists think your body needs a chance to rest and repair itself. Others believe that sleep—rather than just rest—is needed only for your brain, not for your body. It gives your complicated central nervous system a chance to relax. Whatever the reason, you usually feel stronger and healthier after a good night's sleep.

Why are some people left-handed?

Although most people are right-handed, some are left-handed, and a few can use one hand as well as the other. We say these people are ambidextrous (am-beh-DECK-struss). Not all scientists agree on what causes these differences. Many think this is the answer:

Each side of your brain controls the muscles on the opposite side of your body. In most people, the left side of the brain is more powerful than the right side. These people have better control over the muscles on their right side. If the left side of your brain is dominant, or more powerful, you are right-handed. If the right side of your brain is dominant, you are left-handed. If both sides of your brain are about equal, you may be ambidextrous.

How do you taste different flavors?

You actually use your tongue, your nose, and your brain to taste things. Stick out your tongue and look in the mirror. You will see little bumps on your tongue. Inside each of those bumps are about a dozen tiny taste buds. Nerves carry "taste messages" from these taste buds to your brain.

You have four kinds of taste buds on your tongue. The different kinds are in different places. In the back of your tongue you taste bitter foods. You taste sour things on the sides. You taste sweet and salty foods on both sides of your tongue, and also at the very tip.

But that is only the beginning of tasting. The smell of foods also plays a big part in how they taste. That is why foods seem to taste funny and have very little flavor when you have a cold and your nose is all stuffed up.

How does your brain help you smell?

The smell or odor of different things is made by a small amount of gas coming from whatever it is that has the odor. You smell something when the gas touches special nerve cells high up in your nose. They send a "smell message" to your brain. Some flowers, for example, have a weak odor. So if you want to smell them, you sniff to bring the gas up to your smelling nerve cells.

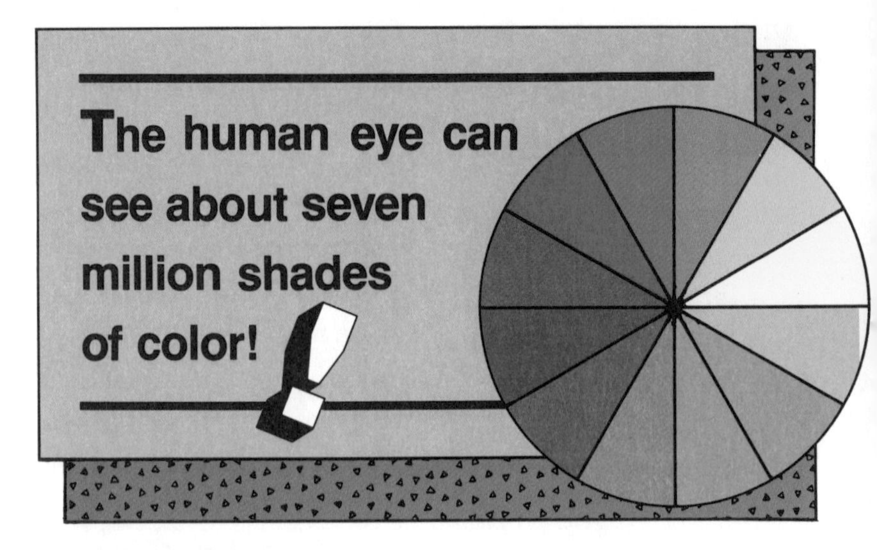

How does your brain help you hear?

When sounds are made, they set up movements in the air. These movements are called sound waves. The outside part of the ear is shaped like a shell. It collects the sound waves and sends them through the inside parts of your ear to nerve cells. The nerves pass the "message" of the sound waves to your brain—and you hear!

How does your brain help you see?

You see with your eyes, but also with your brain. First, light passes into your eye and forms an upside-down picture on your retina. The retina has special nerve cells on it. When the light hits these cells, they send a "picture message" to your brain. Your brain interprets the message into a right-side-up picture—and you see.

What does it mean to be color-blind?

A color-blind person cannot tell all colors apart. Most color-blind people can see shades of yellow and blue pretty well, but red and green look alike to them. A few color-blind people cannot see any colors. They see everything in black, white, and shades of gray. This condition is inherited, but it is not common. More boys are color-blind than girls.

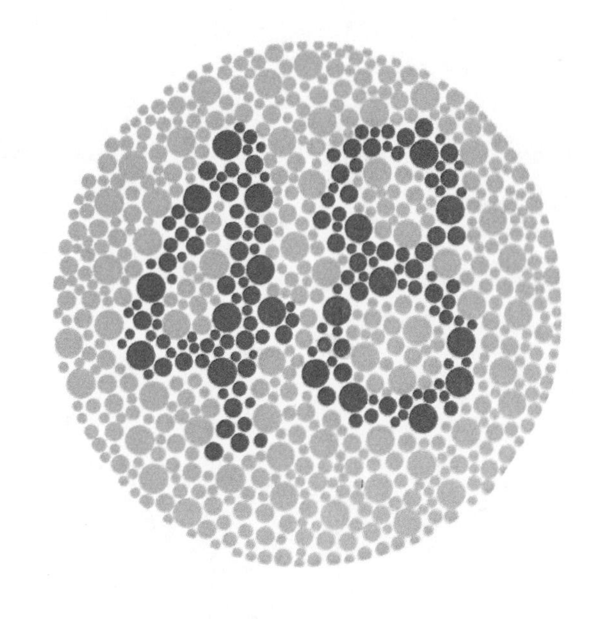

This is a color-blind test.
Can you see the number 48?
A color-blind person can't.

CHAPTER · 4

Is mealtime *your* favorite time of day? Eating is fun, especially when you are hungry and have a chance to munch on yummy foods. You also eat to feed your cells so that your body can grow strong.

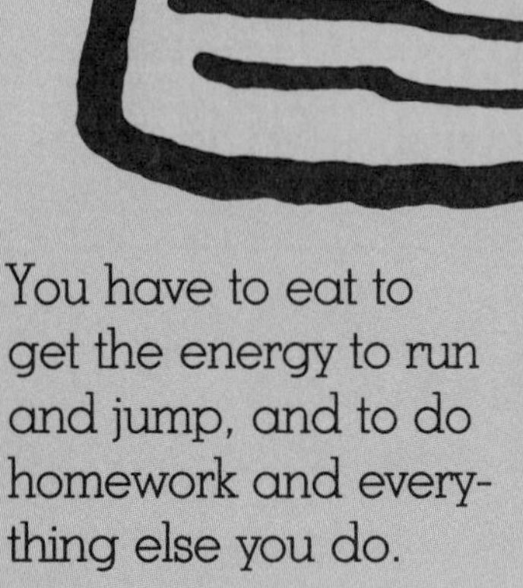

You have to eat to get the energy to run and jump, and to do homework and everything else you do.

FOOD FOR THOUGHT

EATING AND DRINKING

What happens to the food you eat?

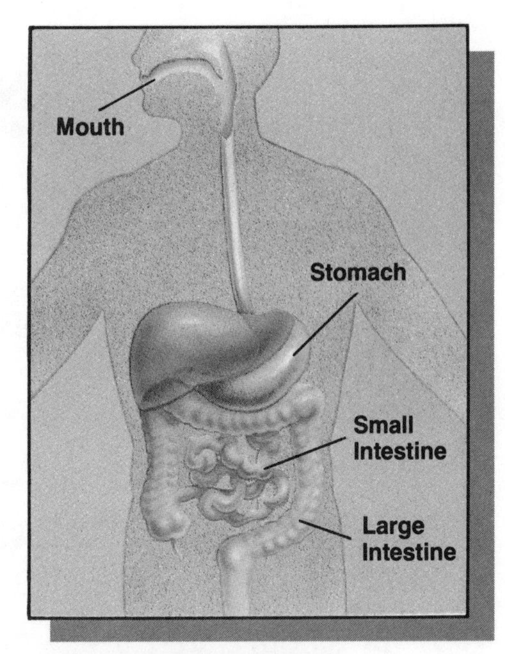

THE DIGESTIVE SYSTEM

You digest (die-JEST) it. That means that your body breaks the food down into pieces small enough to enter your tiny cells.

You start breaking down the food in your mouth. Your teeth chew it into very small pieces. When you swallow the food, it moves down a tube to your stomach. From your stomach it passes through a long, thin coiled tube called the small intestine.

All along the way the food is broken down more and more by juices—digestive juices—that are made in your body. Other chemical changes take place in the stomach. Finally, in your small intestine, most of the food becomes a liquid. The liquid goes into your blood and travels around your body to feed all your cells. The parts of the food that you can't digest continue on into a large, coiled tube called the large intestine. Then the undigested food leaves your body as waste.

Why do you get thirsty?

Your body has a lot of water in it—salt water. You must have certain amounts of both salt and water in your body at all times. When you eat a lot of extra salt, your body has too much salt in it for the amount of water. The same thing is true when you lose a lot of water. The thirsty feeling is a signal to drink more water and get the salt and water levels back to normal.

Why does your mouth water when you smell food?

Your mouth waters because the smell of food starts your digestion going. The "water" that comes into your mouth is not really water at all. It's a digestive juice called saliva (suh-LIE-vuh). There is always some saliva in your mouth. When you eat, a lot more of it flows in to help digest your food.

You don't have to put food into your mouth to start the saliva flowing. You don't have to smell food, either. In fact, you can make your mouth water without even seeing food. Wait until you're very hungry, and then think of your favorite food. Bet your mouth waters!

Why does your stomach rumble when you're hungry?

When you eat, food goes into your stomach. There, a digestive juice called gastric juice helps to break down the food. At the same time, muscles in your stomach start working. They cause the sides of your stomach to move. The movement churns the food and rolls it around to help break it down.

Because you usually have your meals at the same time each day, your stomach gets right to work at those times—even when you haven't eaten. If there is nothing in your stomach, all that churning can sometimes get noisy.

Why do you burp?

You burp to get rid of gas in your stomach. When you eat fast, you swallow a lot of air. Air is a gas. Too much air in your stomach makes you feel uncomfortable. Your body gets rid of it by forcing the air back out through your mouth. If you drink something with a lot of fizz in it, you may also have to burp.

What kind of food should you eat?

If you want to keep your body healthy, you must feed it vitamins, minerals, proteins, fats, sugars, and starches. You can get all of these by eating different kinds of food from the four groups every day. The four groups are:

1. cereal, pasta, grains, and bread
2. fruits and vegetables
3. milk and yogurt
4. fish, poultry, lean meat, cheese, and eggs

VITAMINS

What are vitamins?

Vitamins are special elements in certain foods. They help your body change the food into energy and help keep your body in good condition. Vitamins are not foods themselves, but they are important. They help you thrive and grow! If you eat a balanced diet of foods from the four food groups, you will get all the vitamins you need.

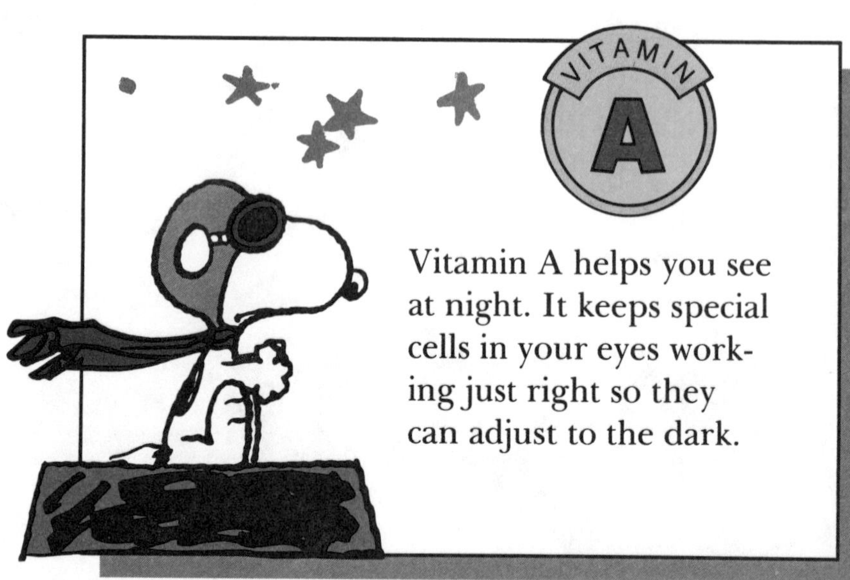

Vitamin A helps you see at night. It keeps special cells in your eyes working just right so they can adjust to the dark.

VITAMIN B group

There are at least 12 different vitamins that make up the B group. Some of them help release energy from food. Others keep your blood and nerves healthy. And some B's make your hair grow!

Vitamin C helps keep your skin, blood, and bones healthy. It also helps wounds heal. Hundreds of years ago, many people died because they couldn't get enough fresh fruit and vegetables, which have vitamin C. Nowadays we have plenty of those foods to help keep our bodies strong.

Vitamin D is the sunlight vitamin! Your body makes vitamin D when rays of sunshine fall on your skin. (But don't forget to use sunscreen.) You need lots of this vitamin so your bones will grow strong and hard. It also helps build strong teeth.

Vitamin E helps your blood carry oxygen around your body.

Vitamin K is the "bandage" vitamin. When the bleeding stops after you cut yourself, that's vitamin K at work!

VITAMIN	WHAT FOODS HAVE THESE VITAMINS?
A	Fish oils; milk and dairy products; cheese; eggs; margarine; leafy green vegetables; yellow fruits such as apricots; carrots; tomatoes.
B group	Milk and dairy products; lean meat; fish; wheat germ; whole grain cereals; nuts; cheese; green peppers; broccoli and other green vegetables; potatoes; nuts; poultry.
C	Citrus fruits such as oranges and grapefruit; green vegetables; tomatoes; potatoes.
D	Milk and dairy products; margarine; eggs; fish liver oils; oily fish. Also produced by sunlight on skin.
E	Seeds; leafy green vegetables; nuts; whole-wheat bread; margarine; cereals; egg yolk; vegetable oil; wheat germ.
K	Green vegetables; cauliflower; liver; oils; cereals; fruit; nuts.

Orange juice is full of vitamin C. A crabby, lazy person can sometimes become a happy, energetic person just by drinking some orange juice every day!

What are calories?

Calories are measurements of the amount of energy provided by different foods. People get overweight when they take in more calories, or energy, than they use up in daily activities. The extra energy is stored in the body in the form of fat.

Everything you do uses up calories. If you walk for an hour, you can use up 300 calories—and burn off a whole hamburger!

Some foods give you more energy than others for the same number of calories. For example, a medium-sized banana has about the same number of calories as a tablespoon of mayonnaise. But the banana has much more good food value.

What is cholesterol?

Cholesterol is a type of fat found in some foods, such as red meat and eggs. Too much cholesterol can be bad for you. It can build up in your body and block the flow of your blood.

CHAPTER · 5

What do you do every day but probably *never* think about? Take a deep breath. Now let it out. Know the answer now? It's breathing!

TAKE A DEEP BREATH

BREATHING IN AND OUT

Why do you breathe?

You breathe to stay alive. When you take a breath, you take air into your body. In the air is a gas called oxygen that your body must have. Oxygen changes the food you have eaten into energy. Your body uses the energy to keep you warm, make new cells, move your muscles, and send messages along your nerves throughout your body.

What happens when you breathe?

When you breathe in, air goes through breathing passages in your nose or mouth. From there the air goes into a tube called your windpipe, then down into your lungs. There, oxygen is taken from the air and passed into your bloodstream. Your blood carries the oxygen around your body to all cells.

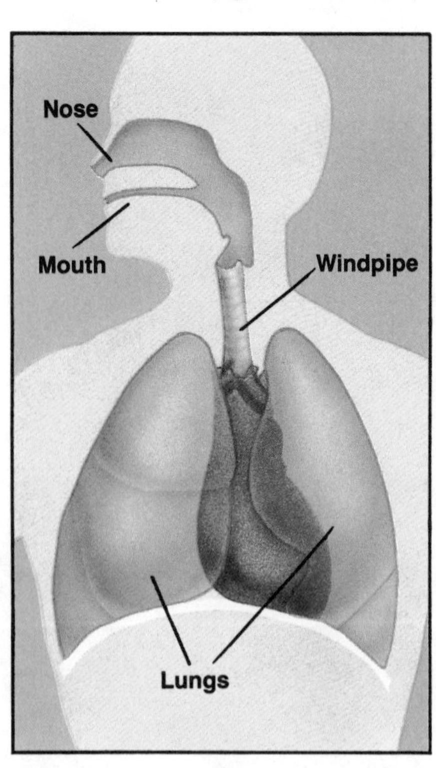

At the same time, your blood picks up a waste gas called carbon dioxide from all your cells. Your blood carries the carbon dioxide to your lungs. When you breathe out, you get rid of this carbon dioxide. Muscles between your ribs and under your lungs tighten and relax to pump these gases in and out.

How do you talk?

Put your fingers on your throat and say, "Yes." You'll get a buzzing feeling on your fingers. The buzzing comes from the "voice box"—or larynx—inside your throat.

Inside your larynx are two vocal cords. When you speak, air comes from your lungs and passes between the vocal cords. The movement of the air makes the vocal cords vibrate—move back and forth very quickly. The vibration causes sound. You use your lips, tongue, teeth, and face muscles to form words.

Why do boys' voices "change" or get deeper when they become teenagers?

Whether a voice is high or low depends on the length of the vocal cords. The shorter the cord, the higher the voice. Girls and women have shorter cords and higher voices than men. When a boy becomes a teenager, his vocal cords lengthen and thicken quite suddenly, causing his voice to deepen. Sometimes it "breaks" abruptly because he hasn't yet learned to master the muscles controlling the vocal cords. This is perfectly normal.

Why do you sneeze?

You sneeze to get rid of something that is bothering one of your breathing passages. You may have dust or pollen—the powder that comes from flowers—up your nose. A message goes to your spinal cord saying, "Get rid of it!" Your spinal cord then sends a message to your breathing muscles. They tighten and relax to make you suddenly breathe in and out with a lot of force—*a-choo!* Out goes the dust or the pollen from your nose.

When you have a cold, you sneeze because cold viruses irritate your breathing passages, making them sore and itchy. But be sure to cover your mouth when you sneeze.

Sneezes travel amazingly fast. Some have been recorded at more than 100 miles per hour!

Why do you hiccup?

You hiccup because of a disturbance in the nerves that control the breathing muscle at the bottom of your chest. Nobody knows for sure what causes hiccups, but most people get them at some time in their life. Some babies get them before they're born.

Many cures have been proposed for hiccups. Some people say taking a deep breath and holding it as long as you can will work. You could also try drinking a big glass of water very quickly, taking tiny sips. Whatever you do, or even if you do nothing, hiccups usually go away quickly.

Why can't you hold your breath for more than a few minutes?

Your body has a built-in protection against holding your breath a long time. When you stop breathing, you begin to store up carbon dioxide. If there is too much of this gas in your blood, a message goes to the part of your brain that controls breathing. Your brain sends back a message to your breathing muscles to start working. Soon you are forced to breathe again, no matter how hard you try to hold your breath.

Why can't you breathe underwater?

Your body needs to breathe oxygen. There is oxygen in water, but your lungs are not built to separate it from the water. They can take oxygen only from air. If you tried to breathe underwater, water would fill up your lungs, and you would drown.

CHAPTER · 6

THE HEART OF YOU

Make a fist with your hand. That's about the size of your heart. That's about the shape of your heart, too. It doesn't look as pretty as a valentine heart, but it's much more valuable. It pumps blood through your body, and that's what keeps you alive.

YOUR HEART

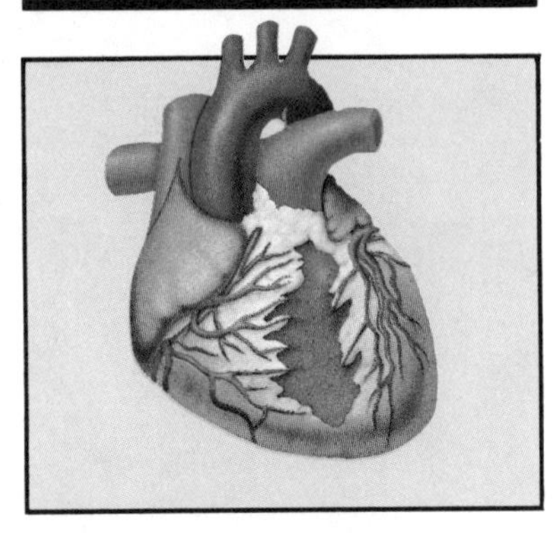

What does your doctor hear through a stethoscope?

Lubb-dub, lubb-dub, lubb-dub. That's the steady rhythm your doctor hears through a stethoscope when listening to your heart. The heart is a muscle that is constantly tightening and relaxing as it pumps blood. We call this constant movement your heartbeat. The *lubb-dubs* are the sound of the strong valves of your heart opening and closing as your heart beats. These valves act like one-way doors, letting the blood in or out of the heart. The doctor can tell by the sound of the *lubb-dubs* if your heart is working properly.

How fast does your heart beat?

Normally, your heart beats about 70 to 80 times a minute. When you exercise, your heartbeat is faster. When you sleep, your heartbeat is slower.

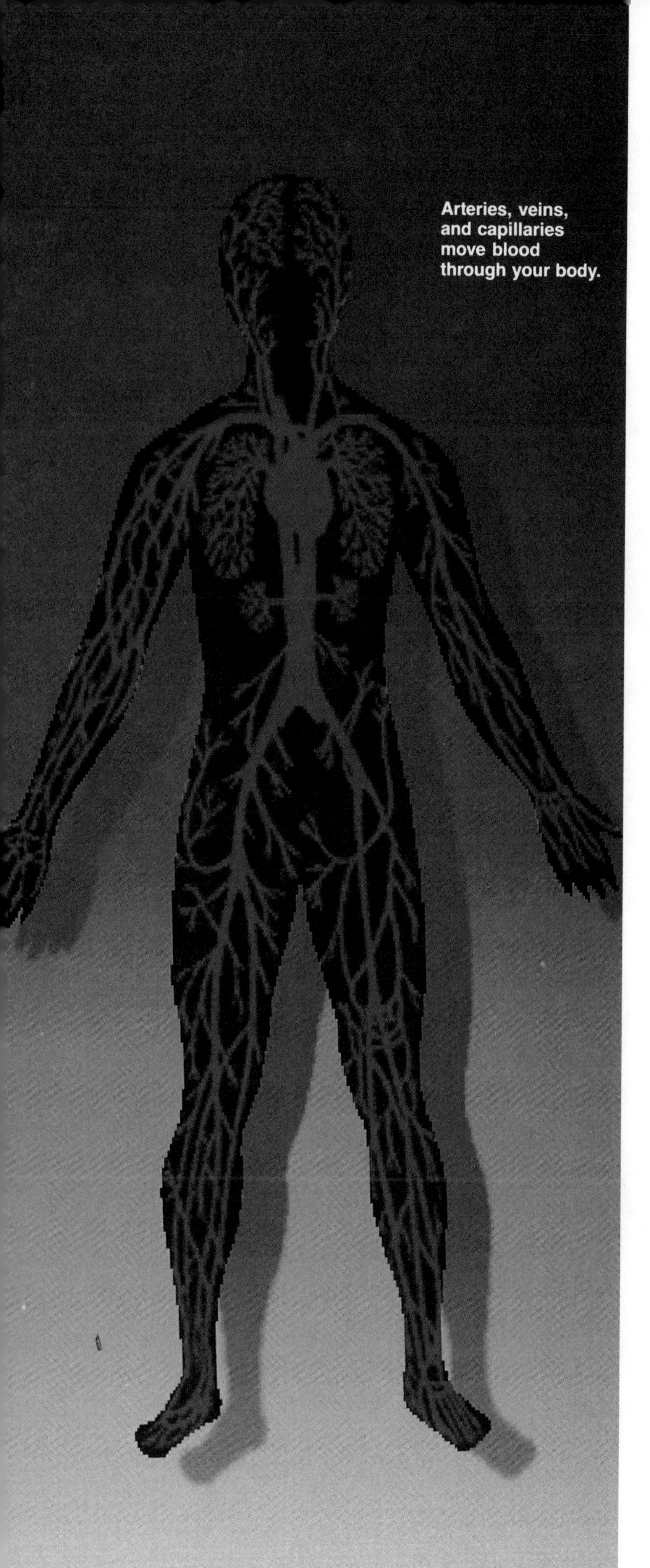

Arteries, veins, and capillaries move blood through your body.

YOUR BLOOD

How does blood travel around your body?

Your blood makes a round trip through your body in less than a minute, thousands of times a day. It travels through tubes called blood vessels. There are three main kinds of blood vessels—arteries (ARE-tuh-reez), veins (VANES), and capillaries (CAP-ih-ler-eez). Blood goes into your heart through veins, and out through arteries.

When blood is pumped out of your heart, it goes into large arteries. These branch into smaller arteries, which branch into still smaller ones. The blood flows from the smallest arteries into capillaries.

Capillaries are the tiniest blood vessels. Blood travels from the capillaries into tiny veins. These lead to larger veins. Finally, the largest veins take the blood back to your heart.

Why do you need blood?

Your heart pumps blood throughout your body to feed your cells, clean them, and keep them healthy. Your blood carries food and oxygen to every cell of your body as it passes through the thin walls of your capillaries.

Your heart sends blood to the lungs to pick up oxygen, and then pumps the blood through all your arteries. Nourishment from food enters the blood through the vessels in the stomach and intestines. As the blood brings food and oxygen to the cells, it picks up body wastes and carries them back through the veins to the heart.

Body wastes called urea (you-REE-uh) and uric acid are carried by the blood to your kidneys. There the wastes mix with water and then leave your body as urine (YOUR-in). The waste gas called carbon dioxide is carried by the blood to your lungs. It leaves your body when you breathe out.

Blood also protects you from infections and disease. It has special cells in it that fight bacteria and viruses.

How do your special blood cells fight disease?

The cells in your blood that fight disease are called white cells. These are like an army for your body. They kill harmful bacteria and viruses that get into or invade your blood. When a large number of these invaders enter your body, the number of white-cell "soldiers" grows. Blood moves to the area where the invaders are, so the white-cell soldiers can attack and kill them. The used white cells and dead invaders form a thick yellow liquid called pus. If the pus is inside a sore on your skin, it may leak out.

Why do you bleed?

A cut bleeds because it has opened some of your arteries, veins, or capillaries. Most cuts don't cause much bleeding because they open only the very small blood vessels. Because these vessels are narrow, blood moves through them very slowly, and comes out of them very slowly, too. If you should ever cut a large vein or artery, you would bleed heavily and you'd need to see your doctor!

Why do people stop bleeding?

Most people are born with a natural protection against losing too much blood. As soon as a cut starts to bleed, your body goes to work to stop it. The blood clots—thickens—and stops flowing. Soon you will see a scab on the cut. The scab is nothing more than dried blood.

If you are bleeding heavily, you can help blood clot faster by pressing on the cut with gauze or by dressing it with a bandage.

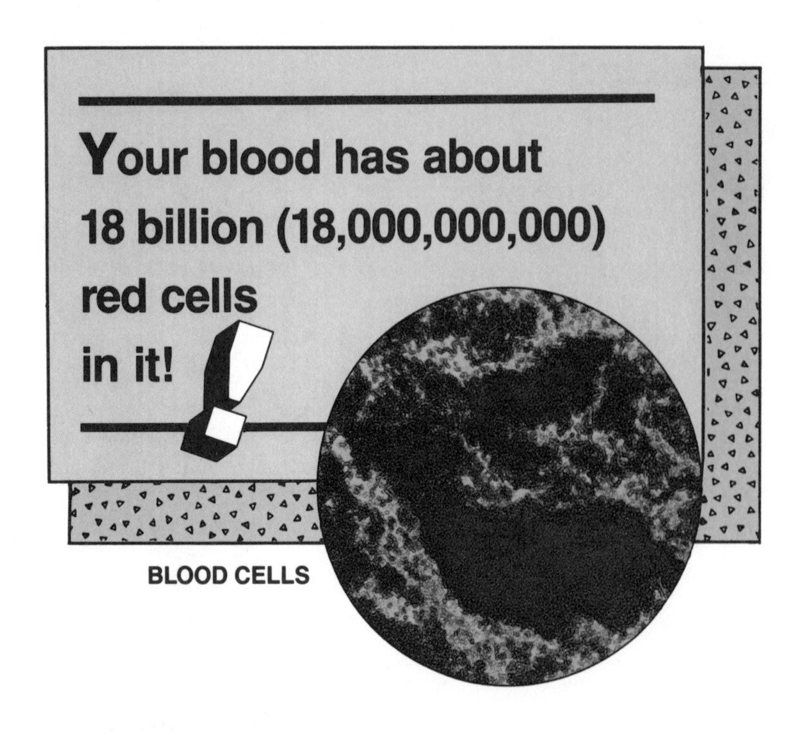

How much blood do you have in your body?

If you weigh about one hundred pounds, you have about seven pounds of blood in your body. That much blood would fill about four quart-size milk containers. If you weigh less, you have less blood. When you become an adult, you will probably have enough blood to fill five or six quart-size containers.

What are black-and-blue marks?

Black-and-blue marks are signs of bleeding under your skin. When you cut yourself, you break open blood vessels and you bleed. When you bump into something, you may also break open blood vessels. Since your skin isn't broken, the blood can't come out. It stays under your skin, where its red coloring changes to yellow, green, and blue. When this happens, we say you have a black-and-blue mark.

Why is blood red?

Blood looks as if it's solid red, but it's not. If you look at blood under a microscope, you will see that it is made up of several different kinds of cells. Only one kind is red, but this one kind gives blood its red color.

What makes your foot "go to sleep"?

Sometimes, after you have been sitting on your foot, you get a prickly feeling in it. That funny feeling means that not enough blood has been moving through your foot. You have been squeezing the veins and arteries so that blood can barely pass through them. Then your blood can't carry the wastes out of your cells, and your nerve cells can't do their job. They aren't able to send messages to your brain, so your foot feels numb. We say it has "gone to sleep." When you get up and stretch your foot out, blood starts flowing again. The nerves in your foot begin to send a lot of messages to your brain. You feel this activity as "pins and needles" pricking your foot.

CHAPTER 7

FEELING GOOD

A*-choo!* Oh, no! You're getting a cold, and you don't feel well at all. Is it a virus? How will you get better? Just listen to Mom and Dad or your doctor, and your amazing body will do the rest.

WHEN YOUR BODY ACHES

What happens when you get sick?

When you feel sick, it's usually because of bacteria (back-TEER-ee-uh) and viruses. Bacteria are tiny living things, so small that you cannot see them without a microscope. Many kinds of bacteria cause diseases. Bacteria are everywhere around us and in us, but our bodies can often fight them off and not get sick. Viruses are even smaller than bacteria and also cause many diseases. Some bacteria can give you pneumonia or strep throat, and some viruses can give you the common cold, measles, or serious diseases such as AIDS.

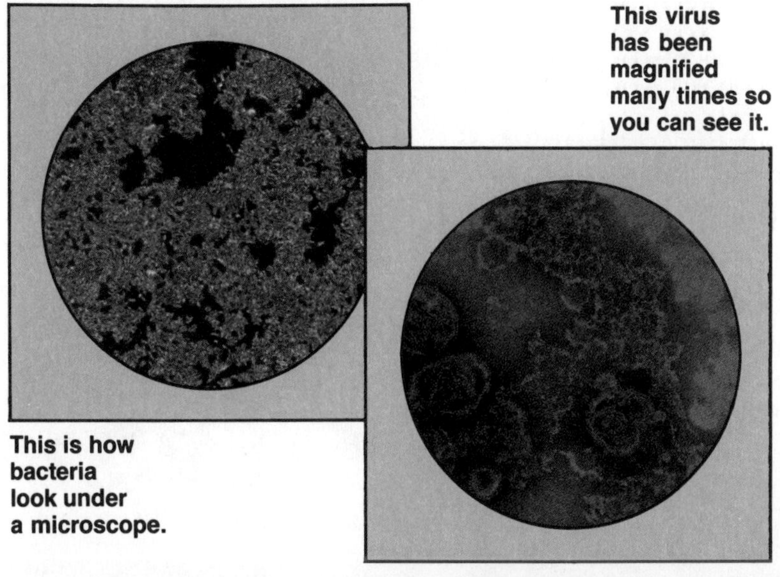

This virus has been magnified many times so you can see it.

This is how bacteria look under a microscope.

Why can't you get the chicken pox more than once?

When you get sick, your white blood cells begin fighting the harmful invaders in your body. One of the ways that the white cells fight is by making special virus killers called antibodies (AN-tee-bah-dees). White cells make antibodies for each particular sickness.

If you get the chicken pox, your white cells make chicken pox antibodies. After you are better, these antibodies stay in your blood and keep killing any chicken pox viruses that get into your body. That is why you can't get chicken pox twice. You have become "immune" to the chicken pox.

What do "shots" do for you?

The doctor gives you your shots or injections (in-JEK-shuns) so that your body will make antibodies for a disease without your ever having to get that disease. When your doctor gives you a shot for measles, he puts a special liquid into your body. It causes your body to make antibodies against measles. Then you become immune to measles.

Sometimes your doctor gives you a shot of "serum." Serum is the part of the blood where the antibodies are. He gives you serum, which already has antibodies in it, to protect you against certain diseases.

If harmful bacteria have made you sick, your doctor may give you a shot of a medicine called an antibiotic. The medicine gets into your blood right away to help you fight sickness quickly. So far scientists have not discovered any medicine that can kill viruses, but your body makes "antibodies" to kill them.

Why can you get many colds?

When you get a cold, your body makes antibodies to fight the particular virus that has come into your body. You will probably never get a cold from that kind of virus again. However, you can have two colds a year throughout your whole life and never have the same cold twice. That's because there are more than 100 different kinds of viruses that cause colds. When one of the new cold viruses enters your body, you will not have antibodies to fight it. Then you will get another cold.

What is a fever?

Fever is a body temperature more than two degrees higher than normal. It is usually a sign that you have an infection or disease somewhere inside you.

Your body is always making heat. Normally, a special part of your brain controls your temperature, keeping it at about 98.6 degrees Fahrenheit. When you get sick, your body heats up to defend itself, since many germs and viruses are killed by heat. So fever helps you fight diseases.

DRUGS AND ALLERGIES

What are drugs?

A drug is any substance you take into your body, besides food, that causes your body to change. Some drugs, such as aspirin or penicillin, are used as medicines to relieve pain or fight illness. They should be used only as recommended by a doctor, because too much of them can be dangerous.

Can drugs harm you?

Yes. There are harmful drugs that people buy illegally. Examples are cocaine, heroin, and crack, a form of cocaine. These drugs are deadly, and should never be taken by anyone.

Other habit-forming or addictive (uh-DICK-tive) substances that can be harmful are nicotine (NICK-uh-teen), found in cigarettes, and alcohol.

What is an allergy?

If you are oversensitive and have a bad reaction to a particular food or substance, you have an allergy (AL-ur-jee) to it. It may give you a headache, an upset stomach, a rash, or some other unpleasant condition.

Some people have a few allergies, others have many. Almost everybody is allergic to some things—to poison ivy, for example.

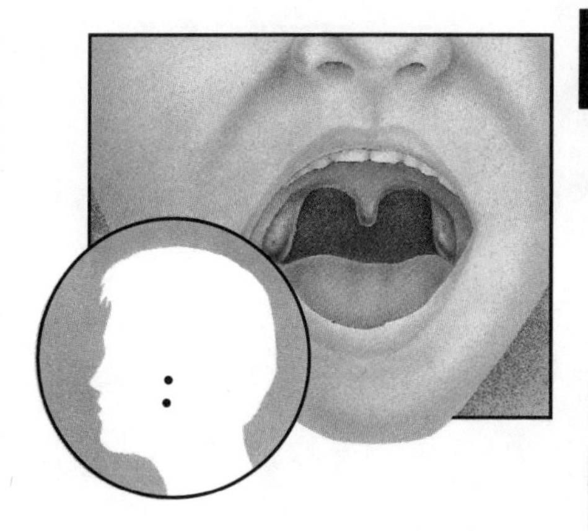

What are tonsils?

Tonsils are small masses of soft flesh in the back of your mouth, at the top of your throat. You have one tonsil on each side of your throat. Tonsils trap harmful bacteria that come in through your mouth, and they make extra white blood cells to fight bacteria.

When you get a sore throat, your tonsils will often swell up. They swell because they are working to help your throat get well. In the past, doctors used to remove the tonsils from children who often had swollen tonsils and sore throats. Now, unless the tonsils become severely infected, they are not removed.

DID YOU KNOW...?

• Everyone has a layer of fat right under the skin. It helps to protect your bones from bumps. Fat even helps you float in the water, since inch for inch, fat weighs less than bones and muscles. The layer of fat also keeps your body warm when the weather is cold, and cool when the weather is hot. If you were on a desert island with no food, your body could use this fat as an emergency food supply. A little fat can help your body a lot, but a lot of fat means it's time to go on a healthier diet.

• Water makes up a big part of your body weight—about three-fifths of it. Even your bones are made up of about one-quarter water.

• If you counted the hairs on your head, they would probably add up to more than 100,000.

• As you grow older, you need less sleep. A newborn baby spends more time asleep than awake. Most adults need about seven to eight hours of sleep a day.

• If you have an average appetite, you'll probably eat more than 35 tons of food in your lifetime. You would need 70 station wagons to bring home all that food.

• In one day of breathing, more than 10,000 quarts of air go in and out of your lungs. That's enough to blow up about 675 beach balls.

• No one in the world has exactly the same voice as you do!

• A "funny bone" is not a bone at all. It's a nerve at the back of your elbow close to the bone. When you hit your funny bone, you get a painful tingling in your arm. But it's not very funny at all!

The ground trembled under their heavy footsteps. Huge and mighty, they roamed the Earth—maybe even in your neighborhood. That was about 200 million to 60 million years ago. Who were these giant creatures? They were reptiles that we have named dinosaurs!

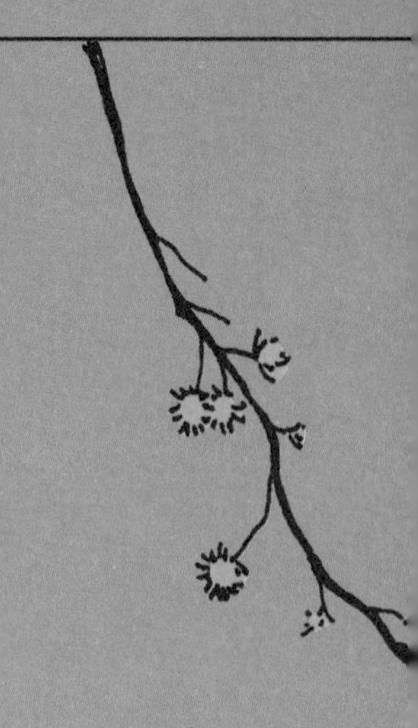

WHEN DINOSAURS WALKED THE EARTH

CHAPTER · 8

THE EARLIEST REPTILES

What is a reptile?

The word *reptile* means that which crawls. A reptile is an animal that crawls, though some prefer to swim. Reptiles usually have scales on their bodies, like fish, but they breathe through lungs, as people do. Reptiles are cold-blooded. This means that the temperature of their blood changes when the air temperature changes. Snakes, turtles, and lizards are all reptiles. So were the ancient dinosaurs.

What does *dinosaur* mean?

The word *dinosaur* means terrible lizard. However, scientists have learned that some dinosaurs looked more like birds than lizards. We think that some of the dinosaurs *were* terrible and fierce. Others were quieter and more peaceful creatures.

Were there people living at the same time as dinosaurs?

No. Dinosaurs died out millions of years before the first humans appeared on Earth. No person has ever seen a living dinosaur. We know how they looked from their bones and from imprints they left in rock.

What were dinosaurs like?

When dinosaurs first appeared, there were not yet any birds or furry animals on Earth. But there were many kinds of dinosaurs in all sorts of shapes and sizes. Some lived on land, and others mostly stayed in the water. Some dinosaurs walked on two legs, and others walked on all four of their legs. There were dinosaurs that ate meat and dinosaurs that ate only plants. Many dinosaurs had a tough plate of armor covering their bodies. This armor helped protect the dinosaurs from enemies.

WHY THE DINOSAURS DISAPPEARED

Why did dinosaurs die out?

No one is sure why dinosaurs died out—became "extinct"—but there are a few possible reasons. One is that the climate of the world changed. The warm, wet places where the dinosaurs lived became drier and cooler. This colder weather killed the plants that some of the dinosaurs ate. When the plants died, the plant-eaters starved. When the plant-eating dinosaurs died, so did the meat-eating dinosaurs—since they ate the plant-eaters.

We know that other kinds of animals appeared on Earth before dinosaurs became extinct. These new animals may have caused the dinosaurs to die out. The animals may have eaten dinosaur eggs. If the eggs were all eaten, there would be no new dinosaurs. Or perhaps the new animals ate the same food as the dinosaurs, and the dinosaurs could no longer find enough to eat.

Another possible reason for the death of the dinosaurs is that they could have been struck down by a deadly, world-wide disease.

COLLECTING DINOSAUR FACTS

How do we collect facts about dinosaurs?

We know how dinosaurs looked, what they ate, how they walked, and many other things—all because we have found their bones and other remains of their bodies. These remains lay buried in the Earth for millions of years and slowly turned to stone. They are called fossils. The word *fossil* means dug up.

Here's a fossil of a dinosaur found in Germany.

The largest number of dinosaur remains in North America have been found in Colorado!

When were the first dinosaur fossils found?

The first dinosaur fossils were found about 175 years ago in Connecticut. Since then, a great many others have been found in other parts of the world. These fossils are mainly dinosaur bones, teeth, and eggs. Scientists can put the bones together into whole skeletons, and from the skeletons they can tell what dinosaurs actually looked like. By studying fossil teeth, scientists can tell whether a dinosaur ate plants or meat. Meat-eaters had pointed, sharp teeth for tearing meat. Dinosaurs that ate plants had flat, blunt teeth, designed for chewing.

Other dinosaur fossils are footprints in the earth that have turned to stone. From these, scientists can tell how a dinosaur walked and how heavy it was.

What colors were dinosaurs?

One thing no one knows about dinosaurs is what color they were. Scientists have found prints of dinosaur skin in stone, but the prints are the color of the fossil stone—not of the dinosaur.

DIFFERENT TYPES OF DINOSAURS

Which were the largest dinosaurs?

When seismosaurus (sighs-mo-SORE-us) was discovered in 1986, some scientists thought that this was the largest dinosaur. It grew to be nearly 150 feet long. More recently, scientists studying fossil bones of different large dinosaurs discovered other dinosaurs, a supersaurus (super-SORE-us) and an ultrasaurus (ultra-SORE-us). Supersaurus probably weighed about 67 tons and ultrasaurus even more. Compare that to an elephant, which weighs about 7½ tons, and a blue whale, which weighs about 100 tons.

You could easily sit in the footprint of a large dinosaur!

How did dinosaurs get such strange names?

The long, hard-to-pronounce names of dinosaurs all come from Latin words. Latin is the language that early scientists used to name living things. When modern scientists discover an animal or plant, they still give it a Latin name. That name is used by scientists all over the world, no matter what language they speak.

When dinosaurs were discovered, scientists gave them Latin names that described what each dinosaur was like. The words *tyrannosaurus rex* (ti-ran-uh-SORE-us recks) means king of the tyrant lizards. Nicknamed *T-rex*, this animal was a fierce, meat-eating dinosaur. The word *brontosaurus* (bron-tuh-SORE-us) means thundering big lizard. This dinosaur, a close relative of seismosaurus, was so big that the ground probably rumbled like thunder when it walked.

Which dinosaur had the smallest brain?

Stegosaurus (steg-uh-SORE-us) had a tiny brain—about the size of a walnut—even though the creature itself weighed nearly 30 tons!

Which dinosaur was the spikiest?

The spikiest of all the dinosaurs was kentrosaurus (ken-tro-SORE-us). This dinosaur had great big horns along its spine, from hips to tail.

Were there any flying dinosaurs?

No. There were no flying dinosaurs, but there were some flying reptiles called pterodactyls (ter-oh-DACK-tilz). None of these reptiles actually flapped their wings and flew like birds. Instead they all glided through the air, sailing along on the wind. Their wings were made of tough skin stretched between their long front legs and short back legs. One flying reptile had a wingspread of 27 feet!

Which dinosaur had its own built-in fan?

Spinosaurus (spy-no-SORE-us) had a huge fan-shaped sail along its spine. Air flowing against the sail helped the animal keep cool.

What other reptiles lived in the days of the dinosaurs?

Quite a few water reptiles were around then. One of these was elasmosaurus (ee-laz-muh-SORE-us). It had a very long neck, and strong legs like flippers for swimming through the water.

Tylosaurus (tie-luh-SORE-us) was a sea reptile that looked something like a modern crocodile. It was a fierce animal with large jaws and very sharp teeth.

Archelon (AR-kuh-lon) was a giant water turtle. The biggest ones each weighed 6,000 pounds and were as long as a large car. Archelon looked very much like any turtle you might see today—except it was much bigger.

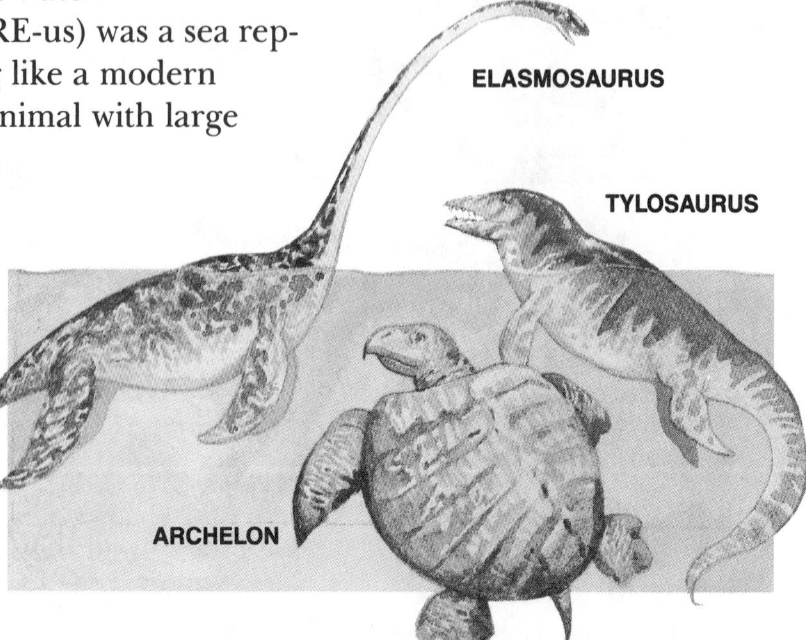

CHAPTER · 9

REPTILES AND SCALY THINGS

Although dinosaurs no longer roam the Earth, there are still lots of reptiles, large and small. Look around. You'll find reptiles in the zoo, or maybe even in your own backyard.

REPTILES

What kinds of reptiles are living today?

Today there are five kinds of reptiles. These are snakes, lizards, turtles, crocodiles and their relatives, and the tuatara (too-uh-TAH-ruh).

Why do reptiles stay underground in winter?

Because reptiles are cold-blooded animals, the temperature of their blood changes with the weather. When the air is warm, their blood is warm, too. When the weather gets cold, the temperature of their blood goes down. The reptiles can get too cold to stay alive. So, to keep from dying, they find a protected place to spend the cold days. They may stay in underground holes, in caves, or under rotting tree stumps. Even in these protected places, the reptiles are too cold to move. They lie still until the air warms up. Then they come outside again. Of course, when reptiles live in places that stay warm all year long, they never have to go underground—except to hide.

What is the tuatara?

The tuatara is a reptile left over from the days of the dinosaurs. All its closest relatives died a very long time ago, but the tuatara somehow survived in one part of the world—on islands near New Zealand.

The tuatara looks like an odd, big-headed lizard. It does everything slowly. It breathes only once an hour. Its eggs take more than a year to hatch, and a baby takes 20 years to grow up.

CROCODILES AND ALLIGATORS

ALLIGATOR

Which is the biggest reptile living today?

The biggest reptile is the saltwater crocodile. This animal is usually about 14 feet long and weighs about 1,000 pounds. Some grow to 20 feet in length—the height of a two-story building!

Do alligators and crocodiles eat people?

Yes, some of them do eat people. Almost any hungry crocodile or alligator may attack a person who comes close to it. But the African crocodile and the saltwater crocodile found in Southeast Asia and Australia are the most dangerous man-eaters. Hundreds of people are killed by these animals every year. American alligators and crocodiles usually leave people alone, though they have been known to attack people.

What is the difference between an alligator and a crocodile?

The easiest way to tell the difference between an alligator and a crocodile is to look at their faces. The crocodile's face is long and pointy. The alligator has a shorter, wider face. When you see a crocodile bite down, its teeth interlock. When an alligator bites, its top teeth come down over its lower jaw.

CROCODILE

What does it mean when you say someone is shedding crocodile tears?

It means that the tears are not true, and the person doesn't feel sad. The saying probably comes from the fact that the shape of a crocodile's jaw makes it look as if it is smiling, even though it isn't. And if you can't trust a crocodile's smile, you also can't believe him if he cries.

SNAKES

How long can a snake grow?

Some anacondas grow to be 25 feet long and 3 feet thick. That's about as long as seven bicycles lined up in a row!

Are snakes slimy?

Snakes are not at all slimy. In fact, their skins are quite dry, and they feel something like leather. People may think a snake is slimy when they see one sitting in the sun. When the sun shines on a snake, its skin looks shiny and almost wet.

Why do snakes stick out their tongues?

Snakes stick out their tongues in order to pick up smells and to feel things. Although many people think that a snake's tongue is a stinger, it is perfectly harmless.

Why do snakes shed their skins?

As a snake grows, its skin gets too small and tight for it, just as your shoes get too tight when your feet grow. So the snake grows a new skin and gets rid of—or sheds—the old one. The snake may do this three or four times a year.

How can a thin snake swallow a fat rat?

An amazing thing about snakes is that they swallow their meals whole. A snake's jawbones are attached very loosely so that its mouth can stretch very wide. The rest of its body can stretch, too, so a very big meal can fit inside. Large snakes can swallow whole rats and sometimes even whole pigs or whole goats! That's quite a meal.

Do snakes ever eat people?

None of the snakes that live in the United States are big enough to eat people. Most snakes eat only insects, mice, and other small animals. But there are two kinds of snakes that occasionally feast on a human being. Pythons (which can be found only in Asia and Africa) and anacondas (found only in South America) are the two man-eaters. These snakes are not poisonous. They kill their prey by wrapping themselves around it and squeezing it to death.

Are fangs like teeth?

Yes, fangs are hollow teeth with a tiny hole at the bottom. All snakes have teeth, but only poisonous snakes have fangs, too. When a fanged snake bites an animal, a poison called venom is forced through the fangs into the victim. A poisonous snake bites small animals in order to kill them for food. A snake bites people and other large animals only if it is scared and wants to protect itself.

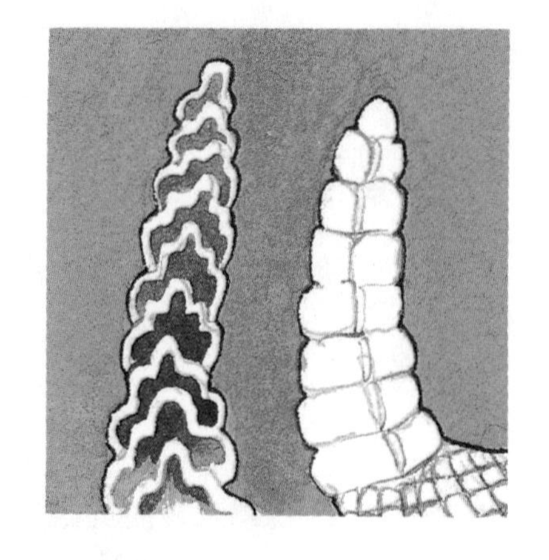

COPPERHEAD

SKULL OF NONPOISONOUS SNAKE

SKULL OF POISONOUS SNAKE SHOWING FANGS

How do rattlesnakes rattle?

At the end of a rattlesnake's tail are a few hard rings made of a material something like your fingernails. When the rattlesnake is excited, it usually shakes its tail. The hard rings hit against one another, making a rattling noise.

Are there poisonous snakes in the United States?

Yes, four kinds of poisonous snakes live in the United States. These are the rattlesnake, the copperhead, the water moccasin, and the coral snake. Of these four, the coral snake has the strongest venom. Fortunately, this snake is small and rarely bites anyone. The other three kinds of poisonous snakes have venom that would take a long time to kill a person. If a person gets bitten, he or she has enough time to go to a doctor and get an antivenom shot.

If you plan to be in an area where you might meet a dangerous snake, you can play it safe. People who hike and work in areas where there are poisonous snakes sometimes carry special medicine in a snakebite kit. You can, too.

Sonora Mountain king snakes look almost exactly like poisonous coral snakes, but they are harmless. Some people keep them as pets!

Can snakes really be charmed?

No. In India, men called snake charmers play music for cobra snakes, and the cobras seem to dance to it. But they are not really dancing. The snakes cannot even hear the music—they are completely deaf! The snakes can feel vibrations in the ground. A snake charmer taps his foot as he plays and sways in time to the music. A cobra feels the tapping, gets excited, and rears up, ready to strike him. When a cobra is ready to strike, it watches its victim carefully and follows the victim's movements. That's just what a cobra does with a snake charmer. The snake charmer is taking a big chance when he excites a cobra. Cobras have a deadly venom, but somehow snake charmers know how to keep an excited cobra from striking. Some snake charmers remove the cobra's fangs to be on the safe side.

HAWKSBILL TURTLE

INDIAN STAR TORTOISE

TURTLES

Is there any difference between a turtle and a tortoise?

Turtles live in the water, but some often come out on land. These are called terrapins. Other turtles—ones with flippers—spend most of their life swimming in the ocean. Tortoises (TORE-tus-uz) are large turtles that always live on land. However, we usually call both creatures turtles.

How long can a turtle live?

No one is sure how long turtles live, but some can probably live for a very long time—100 or maybe even 150 years.

How large can a turtle grow?

The largest turtle is the leatherback. It is a sea turtle that usually weighs between 600 and 800 pounds. The biggest one ever caught weighed nearly 2,000 pounds and was almost eight feet long.

Can a turtle crawl out of its shell?

No. The turtle's shell is attached to some of the turtle's bones.

The matamata turtle has wormy-looking bumps on its neck that fish try to eat. Instead, the matamata eats the fish!

Can you tell the age of a turtle by its shell?

By looking at its shell, you can tell the age of a young turtle, but not of an old turtle. The top of a turtle's shell is divided into sections. These are called shields. On each shield are little circles. In a young turtle, each circle stands for a year's growth. For example, a two-year-old turtle has two circles on each shield. After five or ten years, however, you can no longer find out the turtle's age by the circles. They have either become too crowded together or have begun to wear off.

LIZARDS

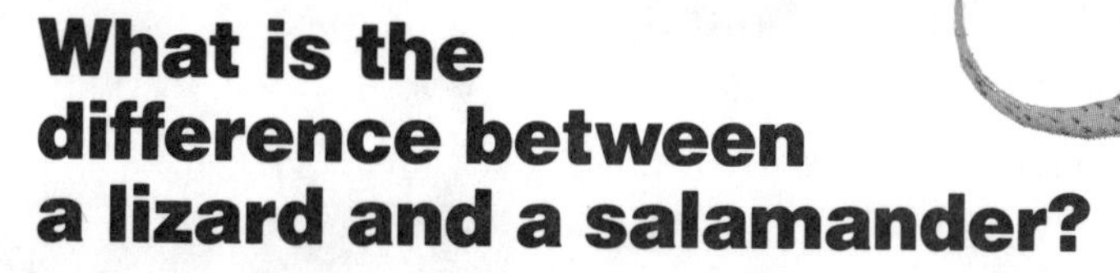

SALAMANDER

What is the difference between a lizard and a salamander?

Some lizards look very much like salamanders. However, the two animals are very different. Salamanders are amphibians. Most amphibians begin their life in water, breathing through gills like fish. Only after they have grown up are they able to live on land. Lizards, which are reptiles, are born with lungs. They always live on land.

Lizards have scales covering their bodies. Salamanders have smooth, moist skin without scales. Lizards love the sun, but salamanders do not. They stay away from it.

Can lizards grow new tails?

Some lizards can. The gecko, the glass snake, and the skink are three of the lizards that grow new tails. If an enemy catches one of them by the tail, the lizard can drop the tail and run away. Then the lizard grows a new tail. If only a piece of its tail is broken off, the lizard will sometimes grow back the missing piece and grow a whole new tail as well. So if you ever see a lizard with two tails, you'll know how it got them.

A gecko can clean its eyes with its tongue!

Are any lizards poisonous?

Only 2 out of about 3,000 known kinds of lizards are poisonous. One of these is the Mexican beaded lizard. The other is the Gila (HEE-luh) monster. The Gila monster lives in Mexico and in the southwestern United States. A bite from one of these two lizards can kill a person, but that rarely happens. The lizards don't usually put enough poison into people to kill them.

Do dragons really exist?

Dragons like the ones in storybooks do not exist. But there are huge reptiles like the creature Marcie is talking to called Komodo dragons. They live on Komodo Island and other islands in Indonesia. They are the largest lizards alive. These dragons can grow to be 10 feet long and weigh 300 pounds. They do not breathe fire, but they do like to eat. So watch out if you're ever near one!

Can lizards change their color?

Some lizards can. These include the anole, sometimes called the American chameleon (kuh-MEE-lee-un), and the true chameleons. They can turn different shades of brown and green. Their color depends on the amount of light hitting them, the temperature, and whether they are calm or scared.

A chameleon often turns the same color as its background to protect itself from enemies. A chameleon on a log, for example, may be brown.

A horned toad—a kind of lizard that lives in the desert—will squirt blood from its eyes when it's scared!

CHAMELEON

Have you ever wanted to soar through the air like a beautiful bird? It dips and circles and sails across the sky, but how does it do it? What's a bird's secret? Wings!

FEATHERED FRIENDS

ALL ABOUT BIRDS

What was the world's first bird?

The first bird was archaeopteryx (ar-ke-OP-ter-ix), which was about the size of a crow. It lived about 140 million years ago and was very much like a reptile. In fact, its ancestors *were* reptiles, and it is thought to be a bridge between reptiles and birds. Like a reptile, archaeopteryx had teeth and a long, bony tail, but archaeopteryx had feathers instead of scales. Also, the wings of the archaeopteryx were much like a modern bird's wings—with bones inside and feathers outside. But archaeopteryx couldn't fly very well. It couldn't flap its wings very hard. It probably used them more for gliding—sailing through the air.

How many different birds are there?

About 9,000 kinds of birds live on the Earth today. Birds can be found almost everywhere except the North and South Poles. The freezing temperatures in the Poles are too cold—even for penguins! There are more birds and different kinds of birds in Africa and South America than anywhere else.

How can you tell for sure which animals are birds and which aren't?

There's only one sure way to tell. See if it has feathers. If an animal has feathers, whether or not it can fly, it's a bird. Ostriches and penguins can't fly, but they are birds. If it hasn't got feathers, it's not a bird, even though it may fly—like bats or insects. All birds have two wings and two feet and no teeth. They have a hard mouth part called a bill or beak, which helps them catch and eat their food.

Why do birds have feathers?

Feathers help a bird keep warm. In cold weather, a bird fluffs up its feathers and traps a layer of warm air under them. The fluffed feathers act like a blanket by holding in body heat. In warmer weather a bird squeezes its feathers against its body to let body heat escape.

Feathers also help a bird fly. In flight, a bird uses its outer wing feathers to move forward in the air. Wing feathers and tail feathers are both used for balancing, steering, and braking.

Do all birds eat worms?

No. Different kinds of birds eat different kinds of food. Usually birds have favorite foods, but will eat some other things, too. Many birds like worms and insects best. Birds that live near water often eat fish or shellfish. Owls, hawks, and eagles eat fish and meat—mice, rabbits, smaller birds, snakes, and other animals. Many small birds, such as sparrows, live on seeds. Some birds eat mostly fruit and berries. Hummingbirds like to drink the sweet liquid called nectar that is found in flowers.

Why do woodpeckers peck at trees?

Woodpeckers peck at trees to get food. They eat insects that live in the trees, just under the bark. Most woodpeckers also peck out nesting holes in trees.

Does an ostrich really stick its head in the sand to hide from an enemy?

No, an ostrich isn't that stupid. What this tall bird does is fall down flat when it sees danger in the distance. An enemy may not spot the ostrich in this position, or it may think the ostrich is just a bush. As soon as danger comes near, however, the ostrich will take off and run. Although an ostrich cannot fly, it can run as fast as 40 miles an hour.

Are owls really wise?

Owls are no wiser than many other birds. In fact, some birds are smarter. But owls have large, staring eyes, which make them look as if they are thinking very hard. That's probably why people started calling them wise.

NESTS AND EGGS

How do birds learn to make nests?

Birds don't learn to build nests. Nest building is an instinct. Each kind of bird is born knowing how to build its own kind of nest. Many birds make a cup-shaped nest out of twigs and grass. Cardinals and thrushes make this kind of nest. Some swallows make their nests in a hole in a tree or rock. They line the bottom of the hole with grass, feathers, fur, and moss. Certain weaverbirds make complicated "apartment-house" nests out of stems. This system of nests may be ten feet high and hold a hundred or more birds.

Do all bird eggs look like chicken eggs?

Most eggs are shaped the same as chicken eggs, but they have different sizes and colors. Large birds lay large eggs, and small birds lay small eggs. The colors of eggs vary from one kind of bird to another. The eggs often blend in with the colors around the nest so an enemy can't spot them easily. Eggs may be light blue, brown, white, gray, or green. A few are red or pinkish orange. Some eggs are spotted or speckled.

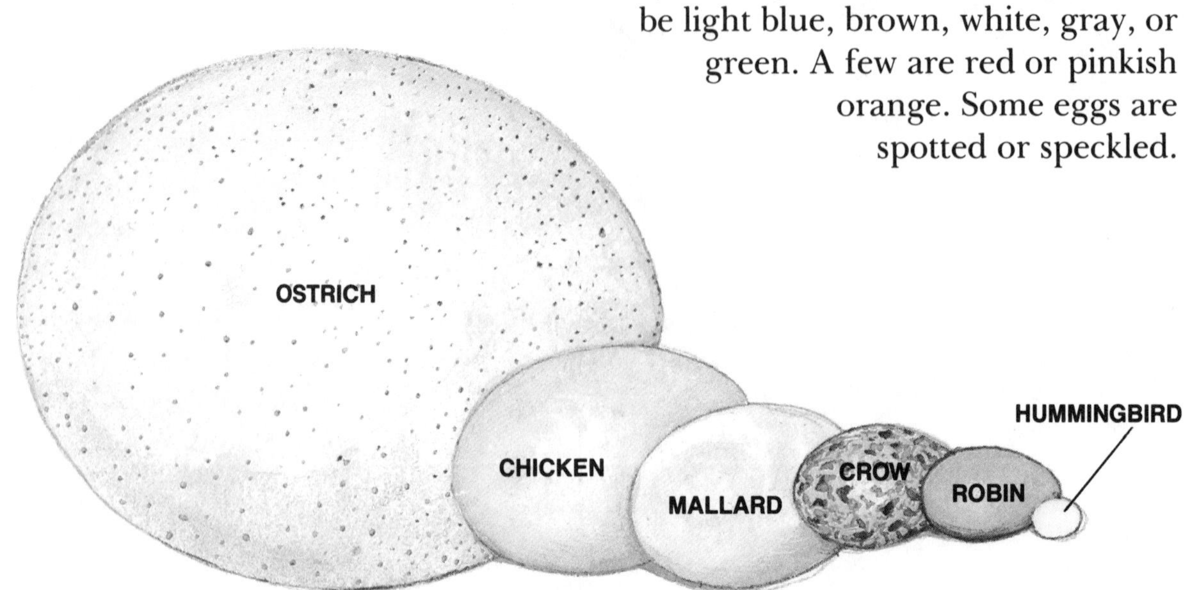

BIRDS IN FLIGHT

Which bird lays the smallest egg?

A hummingbird—which is the smallest bird—lays the smallest egg. Its egg is only about half an inch long.

Which bird lays the biggest egg

The ostrich—which is the biggest bird—lays the biggest egg. This egg can be as long as 8 inches and can weigh up to 4 pounds.

If a 250-pound animal sat down on an ostrich egg, the egg would not break! This is because the egg has a special oval shape that makes it hard to break.

Why do birds sit on their eggs?

Birds sit on their eggs to keep them warm. When an egg is kept warm, the baby bird inside can grow and then hatch, or come out of the shell.

Why can birds fly?

A bird's body is specially built for flying. It is very light. There are pockets of air in it, and most of the bones are hollow. So a bird doesn't have to lift much weight into the air. A bird has very strong muscles for flapping its wings, and the wings have just the right shape for flying. The inner part of a bird's wing is like the wing of an airplane. It lifts the bird up in the air. The outer part of the wing acts like a propeller. Its long feathers pull on the air and move the bird forward. The design of airplane wings was copied from birds' wings.

Are there any birds that can't fly?

The ostrich, the cassowary (KASS-uh-wer-ee), the rhea (REE-uh), the emu (EE-myu), and the kiwi (KEE-wee) are all nonfliers. They have wings, but their flying muscles are not strong enough to be useful. But they are very fast runners. Penguins also can't fly. They have wings like flippers, which they use to swim and dive powerfully. Chickens cannot fly very well, but they can flutter around a bit.

How can a hummingbird stand still in the air?

A hummingbird can stay in one spot in the air, or "hover," because it can beat its wings very fast—from 55 to 90 times in one second! Its wings move so fast that they look like a blur. A hummingbird hovers in front of flowers when it drinks nectar.

Every year Arctic terns fly 11,000 miles south to Antarctica and 11,000 miles back home again. That's 22,000 miles each year!

Where do birds go in winter?

Before winter comes, many birds that live in the north fly south, where the weather is warmer. In the spring, they fly north again. We say that those birds "migrate." No one is sure why birds began migrating, but the need for food was probably the main reason. In cold places there are few insects, flowers, fruits, and seeds around for birds to eat. Ponds and streams are frozen over, so fishing birds cannot get food, either. In warm places, food of all kinds is available.

BIRD SONGS

Why do birds sing?

Bird songs are not just pretty music. Birds usually sing to tell other birds of their kind to keep away from their nesting area. Often birds sing to attract a mate. Sometimes they seem to sing just for the fun of it.

Do all birds sing?

No. Female birds rarely sing, and only about half the males have songs. But nearly all birds give calls. Calls are short, simple sounds. The *whoo-whoo* of an owl is a call. So is the *cluck-cluck* of a hen.

Calls are often used to express alarm and warn other birds of danger. Birds "talk" to their babies with calls. Baby birds use calls to say they are hungry.

PROTECTING BIRDS

How do birds protect themselves?

Birds protect themselves by always listening and watching for danger. At the smallest sign of it, they will fly away. Birds that cannot fly are often able to swim fast, or run quickly and kick, too. Some birds—such as owls—make themselves look dangerous by fluffing out their feathers. Other birds will hiss at enemies and scare them away.

Another important protection for many birds is their color. Their feathers often have colors that match the things around their nest. Some birds, such as the ptarmigan (TAR-muh-gun), change colors with the seasons.

Why have some birds become extinct?

Some kinds of birds have become extinct because people have killed all of them. The dodo, the passenger pigeon, the great auk, and the Carolina parakeet are some of the birds that have become extinct. Hunters have killed birds for their colored feathers, their oil, or their meat. Today some farmers kill larger birds that sometimes eat small farm animals.

People also kill birds without meaning to. When people cut down forests and fill in swamps to build houses and factories, they destroy the homes and the food of birds. If the birds have nowhere else to go and nothing to eat, they die out.

Pollution may soon cause some birds to become extinct. Birds that eat fish from polluted water get poison in their bodies. Then they can't lay healthy eggs. New birds aren't born.

CHAPTER · 11

INCREDIBLE CREATURES FROM LONG AGO

If you could take a peek into the past, you'd see more than just dinosaurs. Other creatures roamed the Earth millions of years ago. They were the world's first mammals.

WHAT IS A MAMMAL?

What is a mammal?

A mammal is an animal that drinks milk from its mother's body when it is a baby. No other animals do this. Most baby mammals grow inside their mother's belly before they are born. Most other animals grow inside eggs that their mother lays.

All mammals are warm-blooded. This means that their body temperature always stays about the same. And they are the only animals that have hair or fur. (Some insects are fuzzy, but they don't have real hair.) Most mammals have four legs, or two arms and two legs.

Dogs are mammals. So are cats, giraffes, bats, cows, horses, rats, monkeys, and dolphins. You are a mammal, too.

THE FIRST MAMMALS ON EARTH

When did the first mammals appear?

The first mammals appeared about 180 million years ago. They probably looked like shrews or rats, having long, pointed snouts and long tails. There were few kinds of mammals on Earth at first, but there were many dinosaurs. As these huge reptiles began to die out about 65 million years ago, many new kinds of mammals appeared on Earth. In a way, these creatures were early versions of dogs, cats, elephants, and horses.

EOHIPPUS

What was the earliest horse?

The ancestor of the horse—the eohippus (ee-o-HIP-us)—was about the size of a small dog. Instead of hoofs, it had three toes on each hind foot and four toes on each front foot. Over millions of years, the horse grew to the size it is now.

Did dogs and cats live on Earth long?

About 25 million years ago, the first doglike and catlike animals appeared. Some of the cats developed into large, fierce animals. One was the saber-toothed tiger. It was about the size of a modern tiger, but two of its front teeth were very long—about eight inches!—and very sharp. Even the largest animals were probably scared of it.

What was a woolly rhino?

One of the earliest mammals was a rhinoceros. It started out small, but as millions of years passed, it became larger. Huge groups of these rhinos moved north to cold lands and grew thick coats of hair. These rhinos were called "woolly" rhinos.

There were also woolly mammoths. They appeared about two million years ago and became extinct only about 10,000 years ago. Mammoths were related to elephants. They were very large and had long, thick hair. Scientists know exactly how they looked, because whole mammoths have been found frozen in ice.

Which was the biggest land mammal that ever lived?

The beast of Baluchistan (buh-loo-chih-STAN). This huge animal looked something like an overgrown rhinoceros. It died out about 20 million years ago. The animal could grow as large as 37 feet long and 25 feet tall. It weighed as much as 22 tons. One of its legs alone was much larger than a grown man!

COLLECTING MAMMAL FACTS

How do we know about early mammals?

We know about them because people have found stone fossils of their bones and teeth in the earth. People have also found real bones and teeth in large pits of tar in La Brea, California. About one million years ago, thousands of animals sank into these tar pits and died. The tar hardened and kept their bones almost perfectly. The bones were very easy to dig out and to study. Saber-toothed tigers, mammoths, vultures, snakes, and camels were some of the animals found in the La Brea tar pits.

Many mammals have also been found frozen in the ice in the far north. Just the way a freezer keeps food from spoiling, the frozen ice kept whole animals from rotting away for hundreds of thousands of years. Many woolly mammoths and woolly rhinos have been found in ice.

A record of animals that lived about 40,000 years ago was left by early people. They painted pictures of animals on cave walls.

CHAPTER 12

Mammals live in almost every part of the Earth—from the coldest to the hottest, the driest to the wettest, on land, in the air, and underwater. There are millions of mammals everywhere—and they do some amazing things!

MEET SOME MAMMALS FROM A TO Z

FOLLOW THE ALPHABET TO SOME OF YOUR FAVORITE MAMMALS

What mammal can live in the coldest climate of all?

The **arctic fox** is comfortable at temperatures as low as 45 degrees below zero. This small white fox seems to be better equipped than even polar bears to manage well in sub-freezing weather.

Are bats really blind?

No. **Bats** can see. In fact, some see very well. However, bats come out mostly at night and many of them have a hard time seeing in the dark. These bats use their ears to help them get around. The bats make little clicking sounds. They can tell by the echo how near or far away an object is. Bats have wings that they can flap, so they are the only mammals that can fly.

How do beavers build dams?

Beavers have four very sharp front teeth. With these teeth, they cut down trees, then cut the trees into pieces. The cut logs and branches are used to make their dams.

A family of beavers usually works together to build a dam. They make a base of logs across a narrow part of a stream. They weigh it down with rocks and branches, then fill the holes with mud. A finished dam is about three or four feet high. It makes a perfect home for beavers, because the deep water around the dam keeps enemies away.

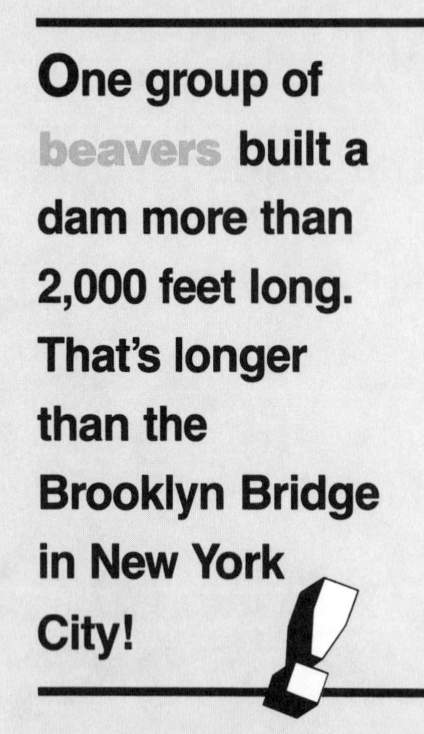
One group of beavers built a dam more than 2,000 feet long. That's longer than the Brooklyn Bridge in New York City!

Do bulls really attack when they see red?

No, they don't. Bullfighters always wave a red cape in front of a bull, but the color red is not what makes the bull charge. In fact, the bull is color-blind. He cannot see red or any other color. But the bull sees the movement of the cape and gets excited.

Why do camels have humps?

Camels live in the desert. They sometimes have to go for a long time without any food. That's when their humps become useful. The humps are made of fat. The camel can get its energy from this fat if it has no food. When the camel has not eaten for a few days, its humps get smaller. They get big again after the camel has filled itself up with food.

Why do a cat's eyes shine at night?

A cat's eyes shine because they reflect light. Even in the darkest night, there is usually some stray light from a streetlamp or the headlights of a car. A cat's eyes reflect this light because they have a special coating on them. The coating helps the cat see in the dark, and also makes the cat's eyes shine.

House cats are not the only cats with eyes that reflect light. Jaguars, lions, tigers, leopards, and all other cats have eyes that shine at night.

What is the world's fastest mammal?

The fastest mammal is a wildcat called the cheetah. It can run at more than 60 miles an hour, and sometimes as fast as 70 miles an hour. But the cheetah can keep up this speed for only a short distance. Then it slows down.

Cougar, puma, panther, painter, mountain lion, catamount, American lion, and Indian devil are all names for the same kind of wildcat!

CHEETAH

Which mammal is most like humans?

Chimpanzees are built a lot like us. They often walk on two feet the way we do, and they have no tail. However, chimps have longer arms, shorter legs, and a more hairy body than we do. The chimpanzee is probably the smartest animal next to man. Some chimpanzees have learned to say and understand a few English words. Others have learned to use the sign language of the deaf. By using sign language, chimpanzees have been taught to answer questions that scientists ask them, solve problems, and even express their feelings.

Chimps also have their own language. They make at least 20 different sounds "talking" to one another.

Why does a cow keep chewing when it isn't eating?

A **cow** has a special stomach with four parts. When it eats some grass, it chews just enough to make it wet. Then the grass goes into the first part of the stomach, where it becomes softer. From there it goes into the second part, where it is made into little balls called "cuds." Later, while the cow is resting, it brings up each cud one at a time and chews it well. When the cow swallows it, the food goes into the third part of the cow's stomach. There the water is squeezed out of it. Finally the food goes to the fourth part of the cow's stomach and is broken down into very tiny pieces. Then the cow's body can take what it needs from the food to live and grow.

Why do dogs pant?

Dogs pant to cool off when they are feeling hot. People cool off by sweating, but dogs don't sweat very much. Instead, they breathe hard, with their tongues hanging out. This brings air into their bodies. The air cools their insides.

Why does a dog wag its tail?

Tail-wagging is one of the ways that a **dog** "talks." You know that a dog is feeling happy when it wags its tail at you. Dogs also use tail wags to give special messages to other dogs. One kind of wag means, "Hello. Glad to see you." Another means, "I'm the boss around here." A third means, "Okay, you're the boss."

How does an elephant use its trunk?

An **elephant** uses a trunk as a nose, hand, and arm. The elephant uses its trunk to smell, to feel along the ground, and to pick up objects. At the tip of its trunk it has either one or two "fingers," which can pick up something as small as a peanut. With its whole trunk, it can lift something as large as a tree!

An elephant also uses its trunk to show affection. A mother pets her baby with her trunk. Both males and females pet each other with their trunks during mating season.

An elephant can also suck up water with its trunk. It drinks by spraying the water into its mouth. Sometimes it sprays water all over its back. This shower keeps the elephant cool and clean.

What is the tallest mammal?

With its long neck and long legs, the giraffe is the tallest animal in the world. Its head may be 19 feet above the ground. The giraffe's height helps it in two ways. First, the giraffe can easily see a great distance over the flat open land where it lives. If a hungry lion is anywhere near, the giraffe will spot it soon enough to run away. Second, the giraffe can eat the leaves high up on trees. Other animals cannot reach these leaves, so the giraffe doesn't have to worry about missing out on a good meal!

Can groundhogs really predict weather?

No, they can't. Groundhogs, also known as woodchucks, hibernate all winter in a hole in the ground. The story goes that on February 2—Groundhog Day—the groundhog comes up out of its hole. If the day is cloudy and the groundhog can't see its shadow, the cold days of winter are over. If the groundhog sees its shadow, the animal returns to its hole. Then we are supposed to have six more weeks of cold weather.

The story is fun, but there is no truth to it. Groundhogs stay in their holes until the weather warms up enough for them to come out. This may happen much later than February 2, or even earlier. Once outside, groundhogs don't look for shadows. They just go about their business—which is *not* predicting the weather!

Why does a kangaroo have a pouch?

A female kangaroo has a pouch so that her baby will have a place to live. When a kangaroo is born, it is only about an inch long—skinny, hairless, and very helpless. It is not ready to live in the outside world. So it crawls across its mother's body and into her pouch. There it can keep warm and safe and drink its mother's milk.

The baby kangaroo stays completely inside its mother's pouch for about six months. Then it begins to stick its head out to eat leaves from low branches. When the baby kangaroo is about eight months old, it jumps out of its mother's pouch for good. Although the young kangaroo is big enough to walk around, its mother still watches it. She pulls it back into the pouch when danger is near.

Is a pony a baby horse?

No, a baby horse is called a foal. A **pony** is a kind of horse that just happens to be small. When fully grown, it is between 32 and 58 inches tall. It weighs less than 800 pounds. That doesn't seem very small until you compare a pony to other horses. A large workhorse can weigh more than 2,000 pounds!

Can porcupines shoot their quills?

No, but their quills are sometimes found stuck in other animals. That's probably how the "shooting" story got started. Actually, **porcupine** quills come off the porcupine very easily. Its tail is particularly full of loose quills. When another animal attacks, the porcupine swings its tail at the enemy. Quills are driven deep into the enemy's flesh. The enemy runs off in pain. Animals that attack porcupines learn their lesson very quickly and don't bother porcupines ever again.

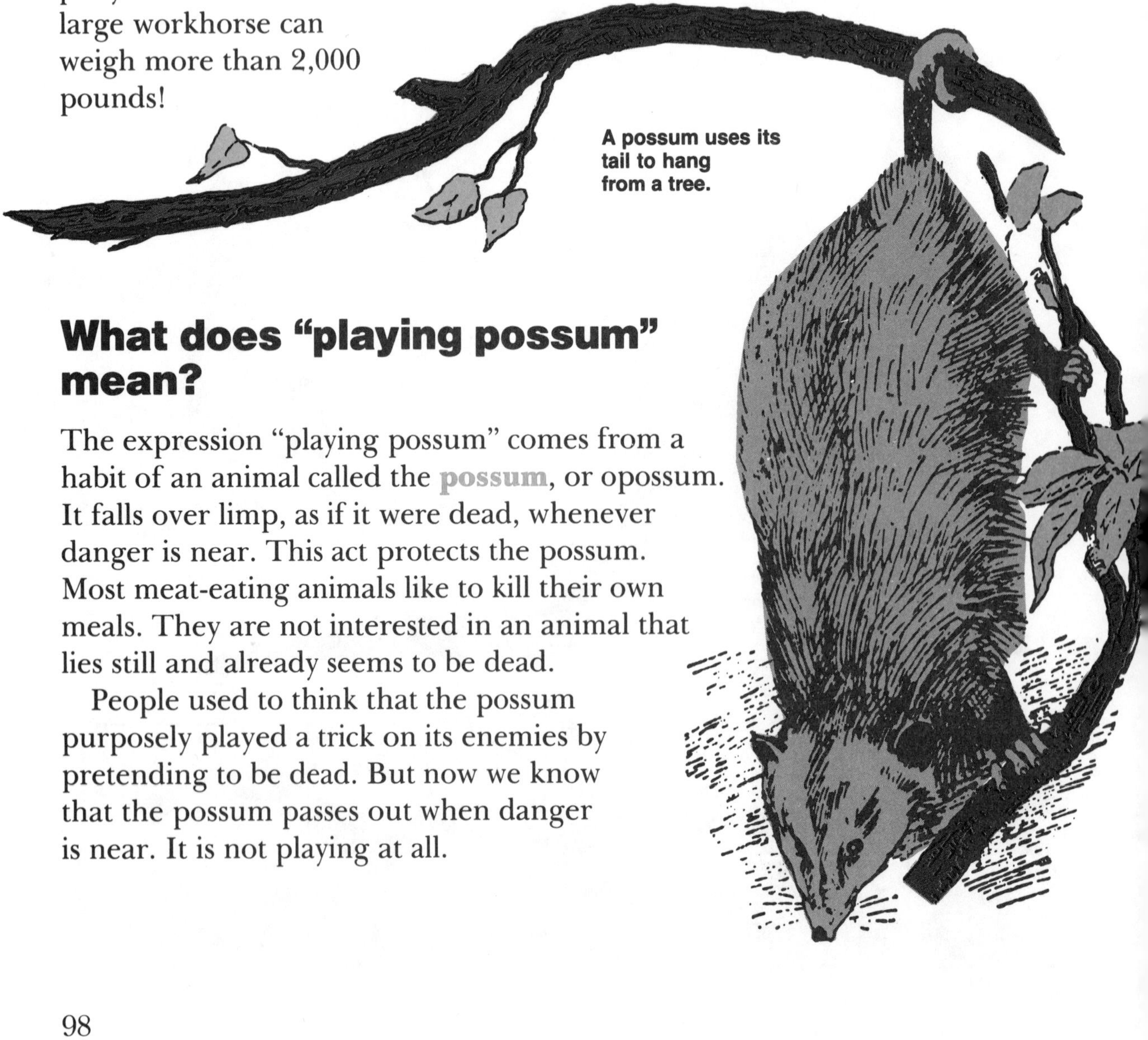

A possum uses its tail to hang from a tree.

What does "playing possum" mean?

The expression "playing possum" comes from a habit of an animal called the **possum**, or opossum. It falls over limp, as if it were dead, whenever danger is near. This act protects the possum. Most meat-eating animals like to kill their own meals. They are not interested in an animal that lies still and already seems to be dead.

People used to think that the possum purposely played a trick on its enemies by pretending to be dead. But now we know that the possum passes out when danger is near. It is not playing at all.

What is the smallest mammal?

The smallest mammal is the **pygmy** (PIG-mee) **shrew**, a small, mouselike creature, which when full grown weighs about half an ounce. These little animals are big eaters, though. A shrew can eat up to three times its body weight every day.

Why do skunks give off a bad smell?

Skunks give off a bad smell to protect themselves from enemies. When a skunk is angry or frightened, it shoots an oily spray into the air. This bad-smelling spray comes from two openings near the skunk's tail. If the spray hits the face of an animal, it burns and stings. It also tastes terrible.

When the spotted skunk gets ready to spray, it stands on its front legs with its back ones in the air!

What is the largest mammal?

The blue **whale** is the largest mammal that has ever lived. It can reach more than 100 feet in length (the height of a 10-story building) and weighs almost 150 tons. That's more than 300,000 pounds. What does this giant whale eat? Small shrimplike creatures called krill.

Why is a whale called a mammal?

A **whale** lives in water and has a fishlike shape and no legs. But a whale is not a fish. It is a mammal, and it acts like one. A whale—like other mammals—grows inside its mother, is born alive, drinks milk from its mother's body, breathes air through lungs, and is warm-blooded. Like all other mammals, the whale has some hair and no scales. A fish, on the other hand, usually hatches from an egg, does not drink milk, breathes underwater through gills, is cold-blooded, and usually has scales.

Why does a zebra have stripes?

A **zebra's** stripes help it hide from enemies. When you see a zebra in the zoo, its stripes make it stand out clearly. But normally the zebra lives in places where there is very tall grass. The zebra's stripes blend in when it stands in the shadows of the blades of grass. The perfect hiding place!

CHAPTER · 13

There's nobody else exactly like you. But you do have one thing in common with every person in the world. You're a member of the human race—a marvelous mammal.

WHERE DO YOU FIT IN?

THE FIRST PEOPLE ON EARTH

When did the first people appear on Earth?

What do scientists call us? Human beings like us are called *Homo sapiens*. Our nearest relatives are believed to have existed between 30,000 and 70,000 years ago. They were named Cro-Magnon people.

Scientists say Cro-Magnon people looked very much like us. Some were over six feet tall and had unusually large brains. Many bones and partial skeletons of Cro-Magnon people have been found in caves such as the Lascaux cave in France. The caves also contain tools, sculpture, and paintings that give us an idea of the skills and artistic abilities of Cro-Magnon people. The nearest relatives of Cro-Magnon people are believed to have been Neanderthal people, who originated about 150,000 years ago. Neanderthal people were short, powerfully built, without chins. Still, in spite of their apelike appearance, scientists think Neanderthal people were intelligent.

THE FIRST PEOPLE IN AMERICA

When did people first come to America?

Scientists think that the very first men and women came to this part of the world between 7,000 and 12,000 years ago. They probably started their journey in northern Asia. After crossing a narrow passage in the sea called the Bering Strait, they probably entered the part of North America that is now Alaska.

People have been on Earth about 30,000 to 70,000 years. Compare this to cats, who have been here for about 8,000,000 years!

DID YOU KNOW...?

RHINOCEROS

There are more than 1,000,000 species (kinds) of animals. Insects account for 800,000 species. All the rest of the animals make up the remaining 200,000 species.

Animals in Danger

Some animals have disappeared from the Earth. They are extinct. Many others also are now in danger of dying out.

No California condors exist free in the world. They are all in zoos. The giant panda and the mountain gorilla are in danger of dying out. And because of game hunters, African elephants and black rhinoceroses are almost extinct.

With the help of zoos and international wildlife groups, many people are working hard to protect endangered animals. They would like to keep these creatures alive, healthy, and free.

Be a Mammal Detective

If you study tracks in the ground or snow in your neighborhood, you might be able to find out what animals live there. It will take a little practice, but soon you'll be able to learn the differences among the tracks of a bird,

squirrel, dog, cat, mouse, and rabbit. Skilled trackers can even tell the difference between the tracks of a black bear and those of a grizzly bear.

Happy Birthday, One and All!

Different animals are likely to live different amounts of time. This list shows that some have surprisingly long and short life spans.

SPECIES	LIFE SPAN
• mayfly	a few hours
• rat or mouse	2 to 3 years
• garter snake	5 to 6 years
• rattlesnake	14 to 18 years
• dog and cat	12 to 15 years
• parrot, swan, and goose	50 years
• alligator	50 years
• elephant	60 to 70 years
• human beings	70 to 85 years
• tortoise	100 to 150 years

People used to live about 70 years. Today, thanks to good diet and health care, the average life span has increased. Some people live through their 80s—and some even get to blow out 100 candles on their birthday cake!

Does It Fit the Bill?

You can tell what kind of food a bird eats by looking at its bill. Seed-eating birds usually have short, stubby beaks—just right for cracking open seeds. The woodpecker's larger bill makes it easy to bore into trees and dig out insects. The eagle is a meat-eater. Its hooked beak helps it tear its food into bite-sized pieces. And the pelican uses the large pouch under its beak like a fishing net for scooping up fish.

Name That Scientist

People who study animals are called zoologists (zoe-AHL-uh-jists). Scientists who study certain types of animals have special names. Here are a few:

SCIENTIST	SUBJECT
entomologist (en-tuh-MAHL-uh-jist)	insects
herpetologist (her-pih-TAHL-uh-jist)	reptiles
ichthyologist (ik -thee-AHL-uh-jist)	fish
ornithologist (or-nih-THAHL-uh-jist)	birds

Talking Animals?

Most animals express themselves, but not with words the way people do. They use movements, smells, and sounds. A kitten meows to its mother to let her know that it's hungry. A bird chatters or sings to warn other birds to keep away from its nest. Some scientists think that dolphins may be able to actually talk to one another!

The Nose Knows

Some mother animals find their babies by their smell. When a baby is born, its mother sniffs it and remembers the scent. From then on, whenever the mother wants to find her baby, she will sniff at all the babies until she finds the right one.

Sleepy Time

Many animals sleep all winter. Animals like ground squirrels, woodchucks, and jumping mice can't find food when the weather gets cold. So they eat a lot and grow very fat before winter comes. Then they sleep—or hibernate (HIGH-ber-nate)—inside a deep hole. They can live all winter on fat stored up in their bodies. When spring comes, warm weather and hunger wake up the sleepers. It's time to stretch and find a tasty meal!

Insects live almost everywhere — from the top of the highest mountains to the bottom of the driest deserts. The ocean is the only place you won't find them. Yes, our world is full of insects. In fact, they make up more than half of all living things on the Earth. But don't let that bug you!

INSECTS!

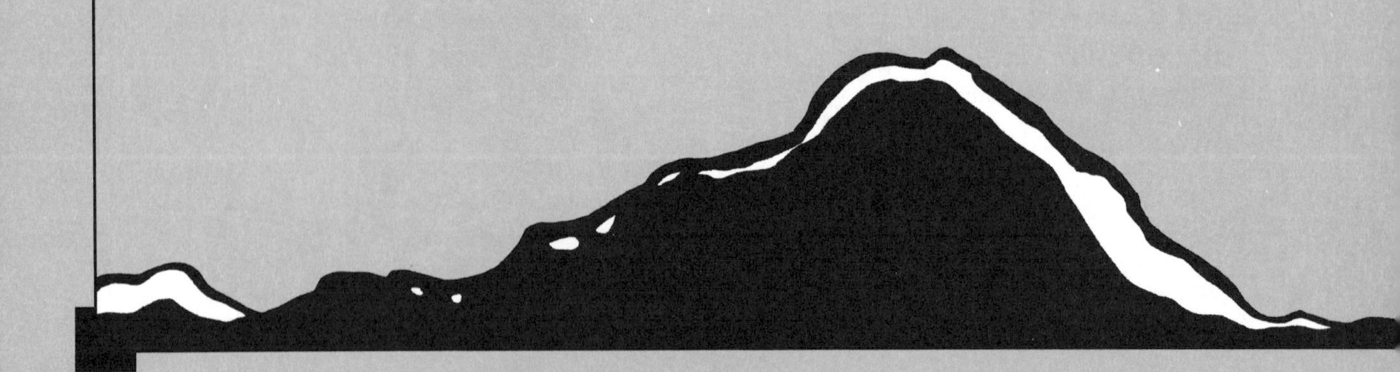

CHAPTER · 14

ALL ABOUT INSECTS

What is an insect?

An insect is a small animal with six legs and a body made of three parts: the head, the thorax, and the abdomen. Many insects have two feelers and four wings, but others don't. There are hundreds of thousands of different kinds of insects, and they all look a little different. Some insects are ants, bees, butterflies, termites, and roaches.

Where do insects come from?

They come from eggs. Female insects lay hundreds or even thousands of eggs during their lives. For example, a queen bee does nothing all summer but lay eggs. On any one day she may lay as many as 1,500. A female termite lays even more. She can lay as many as 30,000 eggs in one day!

If every insect hatched and lived its full life, the world would be overrun by them. There would be no room for anyone or anything else. Fortunately for us, many animals eat insects and insect eggs, so most insects never have a chance to grow up.

Can insects be useful to us?

We make use of insects in many ways. Bees make honey and beeswax. Silkworms make silk. An insect called the lac gives off a sticky liquid that we use to make shellac, a liquid that leaves a hard, protective finish on wood. Other insects help to get rid of harmful ones. For example, the praying mantis and the ladybug eat large numbers of harmful insects.

Bees, butterflies, moths, and other insects carry the yellow dust called pollen from flower to flower. When a flower is pollinated, the plant can grow seeds. These seeds grow to become new plants.

Are some insects harmful?

Yes. There are many harmful insects that spread disease, damage plants, and eat clothing and furniture. And there are insects, such as the mosquito, that bite us.

Ladybugs are useful insects.

Why are insects so small?

Insects are small because of the way they breathe. They have no lungs for breathing air. Instead, they breathe air through tiny holes in their bodies. The air cannot travel very far through these holes. If an insect were larger, air could not reach every part of its body. The insect could not live. So, in order to get air into all parts of their bodies, insects must be relatively small.

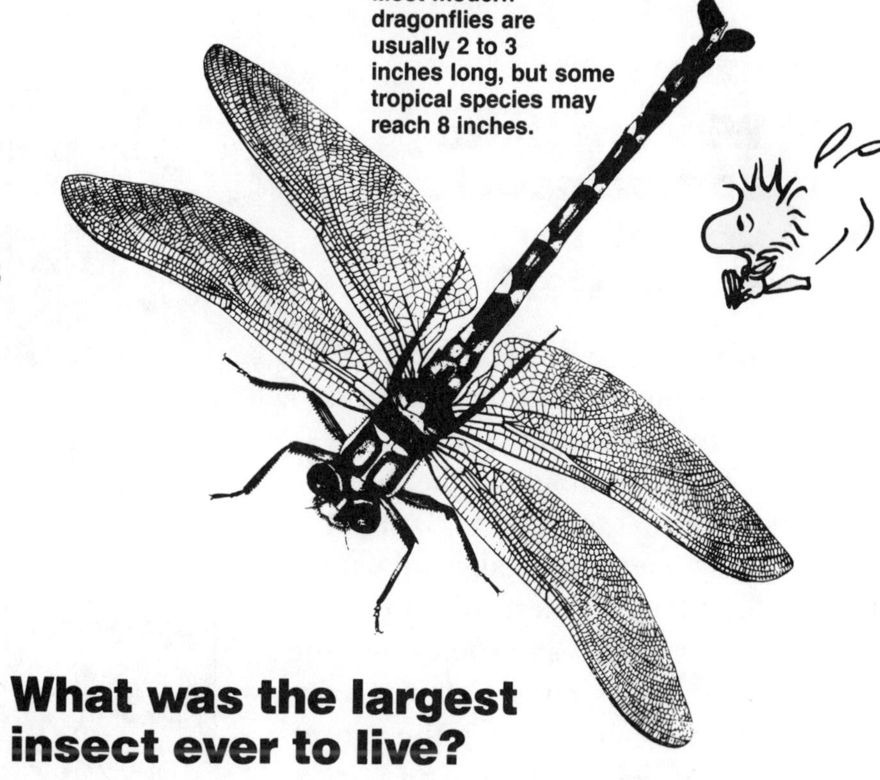

Most modern dragonflies are usually 2 to 3 inches long, but some tropical species may reach 8 inches.

What was the largest insect ever to live?

Many millions of years ago there lived a giant dragonfly whose body was 15 inches long. Its wings measured more than 27 inches from the tip of one wing to the tip of the opposite one. However, this insect's body was only a quarter of an inch thick. If the dragonfly had been fatter, it would not have been able to breathe.

Some Monarch butterflies can travel more than 2,000 miles to the south for the winter!

What is the largest insect living today?

The largest insect is a type of "walking stick" that lives in the tropics. It is very long and thin, and it looks a lot like a twig when it rests on a tree. This walking stick sometimes grows to be nearly 13 inches long.

Why are there so few insects around in the winter?

Most insects die at the end of the summer, but they leave many eggs to hatch in the spring. Bumblebees die, but they don't leave eggs. Instead, their queen stays alive all winter. She sleeps underground until spring. Then she comes out and starts laying eggs. Other insects also stay alive during the winter. These sleep underground or in a barn or cellar for the winter months. Crickets and some beetles do this. Ants do, too, but they come out on warm, sunny winter days. Monarch butterflies are like birds. They fly south to warmer places for the winter.

THE AMAZING ANT

Ants build their tunneled nests underground.

What insects act most like people?

Ants do. They live in nests that are much like cities. Often the nests are built underground and are full of tunnels. They may have roads leading to and from the entrance. Inside the city, ants keep busy doing different jobs. Some clean the tunnels, some take care of babies, and some guard the city. Others go outside and gather food.

There are ants that fight wars. There are even ants that keep other insects as pets. Some kinds of ants grow their own food in gardens. Others keep ant cows.

What is an ant cow?

An ant cow is another name for an insect called an aphid (AY-fid). Aphids make a sweet liquid called honeydew. Certain kinds of ants keep aphids and "milk" them, just as farmers keep cows. An ant uses its feelers to stroke an aphid's sides. The aphid then lets out a drop of honeydew for the ant to drink.

Some ants can carry 60 times their own weight. If you could do that, you would be able to lift approximately two elephants!

BEES AND WASPS

WASP

How do bees make honey?

Only one kind of bee—the honeybee —makes honey. First a honeybee goes to flowers to get nectar. Nectar is a sweet liquid found inside the flowers. A bee drinks the nectar and stores it in its "honey stomach." The honey stomach is not the same stomach that the bee uses to digest its food. It is a special stomach where the nectar is changed into watery honey.

The bee then flies back to its hive. It sucks up the watery honey from its honey stomach and places the honey in little cubbyholes called cells. In the cells, the water dries out of the honey, so the honey becomes thicker.

Why do bees buzz?

The sound of a bee buzzing is nothing more than the sound of its wings moving. So when a bee flies, you hear the buzzzzz.

Do different bees have different jobs?

Yes. Honeybees and bumblebees live in large groups called colonies and divide up the jobs that have to be done to keep the colony alive. Some bees are soldiers that must protect the hive. Other bees called workers go off to collect nectar to make honey.

Why do bees sting?

Bees sting in order to protect themselves from enemies. They do not sting because they are mean. If you don't bother a bee, it will usually not feel threatened by you, and it will not sting you. However, the smell of certain perfumes may cause a bee to sting. So if you are wearing perfume and decide to take a walk outside, watch out!

Does a bee die when it stings you?

Only worker honeybees die when they sting you. No other bees do. Most of the bees that sting have smooth stingers. After one of them stings you, its stinger slips right out of your flesh. The honeybee's stinger, however, has tiny sawteeth at the end of it with a poison sack attached. As the stinger works its way through the skin, poison is also pumped into the skin. When the honeybee flies away, the stinger stays in your flesh. Soft parts of the bee's body pull off with the stinger, and the honeybee soon dies.

How dangerous is the sting of a bee or a wasp?

The sting of a bee or a wasp usually is not dangerous to people. Most of the time the sting hurts a lot, and the area around the sting swells up. After a while, though, the pain goes away, and so does the swelling. Some people, however, are allergic to the sting. They may break out in a rash, or their eyes and lips may swell up. A few people are so allergic to stings that they have trouble breathing and must quickly see a doctor. This extra-strong reaction is not common.

Three hundred babies come out of each egg laid by some wasps!

What are hornets and yellow jackets?

Hornets and yellow jackets are two of the most familiar kinds of wasps. Wasps are related to bees, and are known for their love of fruit juices and for their painful stings. Some kinds of wasps live all alone. Others, including hornets and yellow jackets, live in colonies as honeybees do. All wasps are helpful insects. The adults feed their babies insects that are harmful to people and crops.

What is a wasp's nest made of?

Different kinds of wasps make different kinds of nests. Paper wasps, including hornets and yellow jackets, build their nests of paper. They make the paper by chewing up wood. Some wasps, called mud daubers, make their nests from mud. They build rows of mud cells in protected places, such as under bridges and roofs of buildings. Potter wasps attach their mud nests to plants. These nests look like tiny clay pots! Carpenter wasps dig tunnels in wood for their nests. Digger wasps dig tunnels in the ground.

MOTHS, BUTTERFLIES, AND THE MAGICAL CATERPILLAR

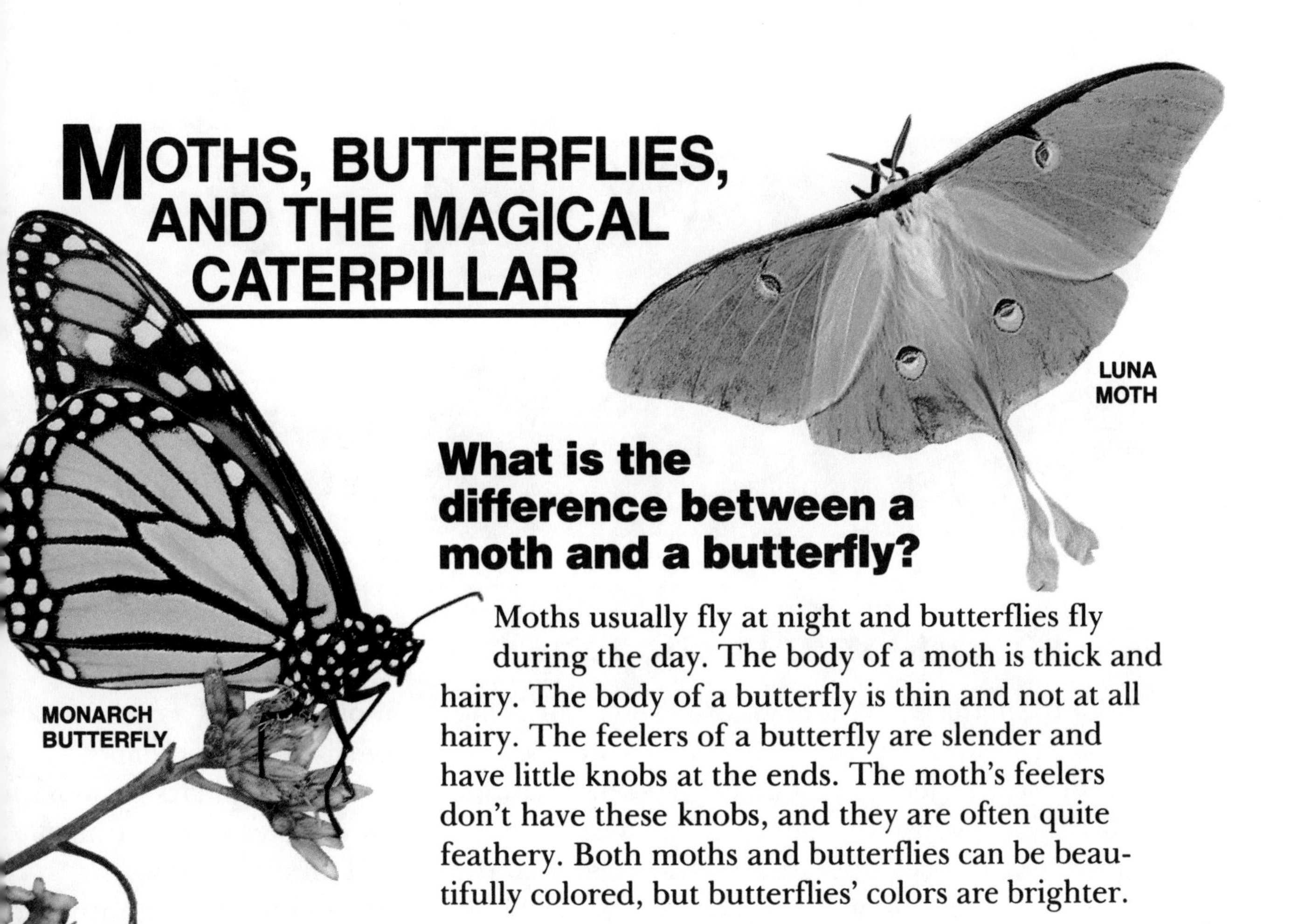

What is the difference between a moth and a butterfly?

Moths usually fly at night and butterflies fly during the day. The body of a moth is thick and hairy. The body of a butterfly is thin and not at all hairy. The feelers of a butterfly are slender and have little knobs at the ends. The moth's feelers don't have these knobs, and they are often quite feathery. Both moths and butterflies can be beautifully colored, but butterflies' colors are brighter.

How does a caterpillar turn into a butterfly?

When a butterfly egg hatches, out comes a worm-like creature called a caterpillar. The caterpillar eats a lot of leaves and grows big. Then it attaches itself to a twig and grows a hard skin. Now it is called a chrysalis (KRIS-uh-lis). For weeks the chrysalis stays very still, but inside the hard covering, many changes are taking place. Four wings, six legs, feelers, and new and different eyes are forming. Finally, the covering splits open. A butterfly with tiny, damp wings comes out. It hangs on a twig until its wings dry out. Then it is ready to fly away.

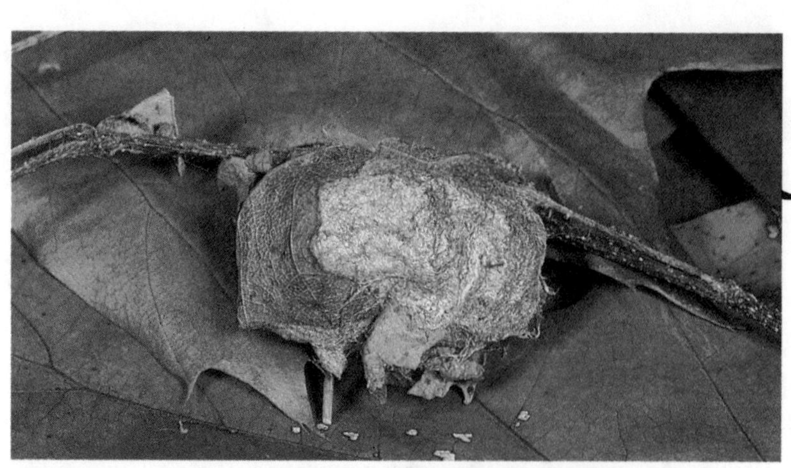

Inside this cocoon, a caterpillar is changing into a moth.

Do butterflies spin cocoons?

No, but moths do. When a moth caterpillar is big enough, it spins a protective case around itself. This case is called a cocoon. The cocoon is spun of silk threads, which the caterpillar makes in its body. The caterpillar rests inside its cocoon and slowly changes into a moth. Like the butterfly, the moth comes out of its covering and flies away.

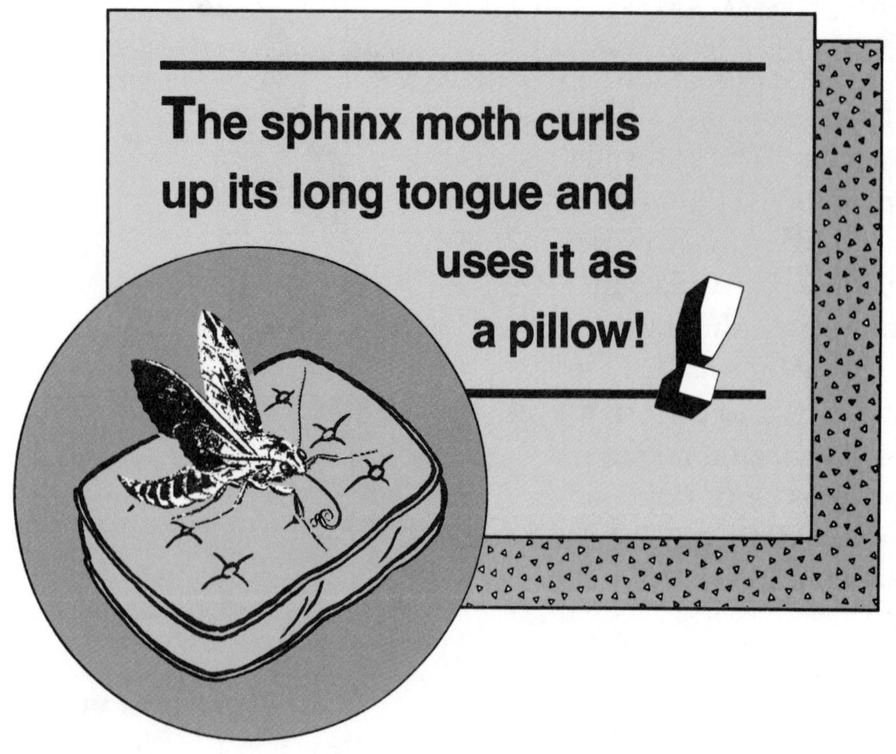

How do we get silk from silkworms?

The silkworm is really a caterpillar that will someday become a small moth. It spins a cocoon of silk just as other moth caterpillars do, but its silk is especially fine.

The silkworm's silk comes out of its mouth as a thread of gluey liquid. The thread hardens as soon as it touches the air. The thread is often as long as 1,000 feet! The caterpillar winds the thread around and around its body to form a cocoon.

To get the silk, people heat the cocoon and kill the caterpillar inside. Next they put the cocoon in warm water to soften the gum that holds the threads in place. Then they can unwind the thread. From the thread, fine silk material is woven.

Why do moths gather around light bulbs at night?

Many insects are attracted to light. They have an instinct to go toward it. A moth is one of these insects. When a light goes on, a moth is drawn to it. The moth can't stay away. Since moths are awake at night, you will often see a group of them flying round and round a light bulb.

Why do moths eat your clothes?

Actually, grown moths don't eat your clothes. Certain moth *caterpillars* eat them. Rugs and clothes are their favorite foods. They especially like wool and fur. These caterpillars get into your closet or drawer if a female moth lays her eggs there. When the eggs hatch, the hungry little caterpillars come out and go to work on your clothes.

SUPER PESTS: FLIES, TERMITES, MOSQUITOES, AND ROACHES

Why are house flies super pests?

Flies are super pests because they can carry bacteria that can make us sick. So we swat them away when they crawl on food—or on us!

How can a fly walk on the ceiling?

A fly can walk upside-down on the ceiling because of the pads on each of its six feet. If you look at a fly with a magnifying glass, you can see these pads clearly. Some scientists think that the fly stays on the ceiling because the pads are sticky. Others believe that the curved pads flatten out against the ceiling and hold on the way suction cups do.

How do worms get into apples?

They are born there! But they're not really worms. They're one stage in the life of insects such as codling moths and fruit flies. For example, in the middle of summer, when apples are growing on apple trees, female fruit flies push their eggs inside some of the apples. The inside of an apple is the perfect place for a young fly because it is moist, protected, and surrounded by food. The eggs hatch into tiny wormlike creatures called larvae (LAR-vee). If no one picks the apples, they fall off the trees in the autumn. The larvae crawl out and bury themselves in the ground. A hard skin forms around each one. Then next summer, a fly comes out of the skin.

Why do termites eat the frames of our houses?

Termites eat wooden house frames because wood is their favorite food—and because wood houses make a perfect home for termites. Termites chew holes that they use as rooms to live in. They line the tunnels and holes with chewed-up wood that they have made into a kind of clay.

Wood-eating termites damage more than the frames of houses. They eat wooden bridges, fences, and boats. In certain countries that are always warm, termites may get *inside* houses and eat furniture, books, and paper. Look what termites did to Snoopy's house!

How do termites digest wood?

Termites get help from tiny creatures called protozoa (pro-tuh-ZOE-uh). Thousands of these one-celled animals live inside each termite. The termite eats wood, and the protozoa digest the wood. The termite is then able to digest what they leave behind. The protozoa and termite are a team. They need each other. In fact, they couldn't survive without each other. When two creatures depend on each other in this way, we call their relationship symbiosis (sim-by-OH-sis).

Do all mosquitoes bite?

No, only female mosquitoes bite. When one bites you, she pricks your skin with a long, thin part of her mouth. Then she sucks some of your blood for food. The mosquito has a special liquid in her mouth to keep your blood thin and easy to suck. Some of this liquid gets under your skin. It causes the bite to swell and itch because most people are allergic to this liquid. However, there are a few people who are not allergic and don't itch at all from mosquito bites!

Will we ever get rid of cockroaches?

Cockroaches are experts at staying alive. They have been around since the days of the dinosaur—many millions of years! Cockroaches can eat almost anything—garbage, soap, book bindings, even television wires! People kill cockroaches with poisons, but cockroach babies are often born immune to the same poison that killed their parents. This means that the babies cannot be killed by that poison, and something new has to be found to do the job.

Cockroaches do not bite, and they usually don't spread disease, but no one wants to have them around. They like damp and dirty places, and places like kitchens where they're likely to find food. So a clean, dry house may discourage them from coming in. But they'll probably still be living out there for the next million years!

WHAT'S IN A NAME?

Do darning needles sew?

From their name it sounds as if they might—and they look as if they might—but they don't. The darning needle's real name is dragonfly. Dragonflies are scary-looking insects, but they are really perfectly harmless. In fact, darning needles are very helpful to us. They eat many insect pests, such as flies and mosquitoes.

Does a doodlebug like to doodle?

No. Doodlebug is just another name for a young ant lion. An ant lion is not an ant and it's not a lion. It is an insect that—to an ant—might seem as ferocious as a lion seems to us.

In the early part of its life, an ant lion digs a pit in sand and buries itself at the bottom. Only its head sticks out. It waits for an ant to fall into the pit. When one does, the ant lion then kills it with its big jaws and sucks the juices out of its body.

How did the praying mantis get its name?

When a praying mantis holds its front legs up together, it looks as if it is praying. However, this insect is not praying at all. It is waiting for a smaller insect to come by so that it can grab the insect with its front legs. The praying mantis will crush the insect and eat it. People like the praying mantis because it eats many insects that harm our crops.

What is the "fire" in the firefly?

Fireflies make two special juices in their bodies. When these juices mix together, fireflies light up. Scientists are not sure why fireflies make this light, but they think that a firefly's glow is probably a signal to attract a mate.

CHAPTER · 15

SPIDERS AND OTHER CRAWLING THINGS

Did you know that spiders are not insects? They belong to the group of animals called arachnids (uh-RAK-nids). This group includes lots of creatures such as scorpions, mites, ticks, and daddy longlegs, too. What's so special about spiders and their crawling relatives? Let's find out!

ALL ABOUT SPIDERS

What's the difference between spiders and insects?

Insects have six legs, and spiders have eight. An insect's body has three main parts. A spider's body has only two. Most insects have feelers and wings. Spiders don't have either.

Are spiders dangerous to people?

Not many spiders are dangerous. Most spiders are harmless to people. In fact, they're very helpful because they eat flies, mosquitoes, ants, and other insect pests.

What happens to spiders in the winter?

Most spiders die when winter begins, but they leave their eggs behind so that baby spiders will be born in the spring. Wolf spiders live longer, usually from 6 to 7 years. Some tarantulas have been known to live for 20 years.

SPIDERS AND THEIR WEBS

How does a spider spin a web?

A spider spins a web out of silk that it makes inside its body. The silk comes out in very thin liquid threads. As soon as a thread touches the air, it hardens. Some of the threads are sticky and some are not. The spider attaches the threads to a tree or house in a particular pattern. One kind of web you may have seen is called an orb web. It looks something like a wheel. Insects get caught in the sticky threads of the "wheel." The spider then kills the insects and eats them.

Do all kinds of spiders spin webs?

No, they don't. Many lie in wait for their prey and then pounce on it. The trap-door spider builds a door that opens into a hole in the ground. It then waits inside for an insect to walk by and fall in.

ORB WEAVER SPIDER

Why aren't spiders caught in their own webs?

A spider is careful to walk only on the nonsticky threads of its web. Even if it does slip and touch the sticky threads, it isn't caught. It is protected by an oily covering on its body.

BLACK WIDOWS AND TARANTULAS

Is the bite of a black widow spider very dangerous?

Yes, it is. But there aren't many cases reported, and there are medicines to treat people who have been bitten.

The full-grown female black widow is the kind that is dangerous. You can recognize it easily because it is shiny black and has a red hourglass-shaped mark on its underside.

Aren't tarantulas dangerous, too?

The tarantula isn't deadly, though it's very big and looks fierce. Its bite is painful, but it doesn't really make a person sick.

WHEW!

POISONOUS BLACK WIDOW SPIDER

DADDY LONGLEGS, SCORPIONS, MITES, AND TICKS

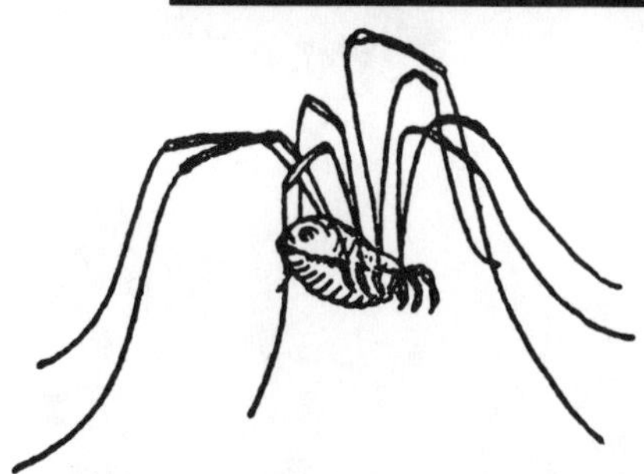

What is a daddy longlegs?

A daddy longlegs is a close relative of the spider, but it does not spin a web. You can easily recognize a daddy longlegs by its tiny body and eight very long, skinny legs. If it loses one of its legs, it will grow a new one! Daddy longlegs don't catch live insects. They eat mainly leaves and grass and insects that are already dead.

Can a scorpion's sting kill you?

Most scorpions aren't really dangerous. Only the large scorpions living in the South have a poisonous sting, powerful enough to make a person very sick.

SCORPION

What are mites and ticks?

Mites are tiny, spiderlike animals that feed mainly on fruits, flowers, and other plant leaves. Some, such as the scabies mite, feed on the skin of people.

Ticks are also very small, but they can carry serious diseases, such as Lyme disease, from animals to people. If a tick sucks the blood of an infected animal and then moves to a person, it can transfer the infection. It's not hard to protect yourself against ticks. When going into the woods, cover your arms and legs, and spray your clothing with a tick repellent. Then when you get home, check your body and clothing for ticks.

There's a living world of wonderful shapes and colors under the sea. Let's go to the ocean floor and visit the creatures that live there. As you read on, Charlie Brown will take you on an exciting underwater adventure!

SOMETHING FISHY

FACTS ABOUT FISH

What is a fish?

A fish is an animal that lives in water and has bones inside its body. Fish are cold-blooded, which means that their body temperature is the same as the water temperature. Almost all fish have fins, which help them swim, and most have scales to protect their bodies.

How many kinds of fish are there?

Altogether there are about 21,000 kinds of fish. We divide these into three main groups. One small group, called lamprey eels, has no jaws at all and suck in food with just their mouths. The second group is very old. These fish don't have real bones inside them but only cartilage (CAR-til-luj), which is softer than bone. This group includes sharks. The third and largest group of fish all have bones inside their bodies. Goldfish, bullheads, and tuna are some of the fish in this group.

Different fish can look amazingly different. Fish are every color you can imagine—red, green, gray, yellow, purple, orange, blue, and brown. Some have stripes, some have spots, others have fancy patterns. Fish vary in size and shape from short and fat to long and thin.

Can any fish live out of water?

Yes, a few fish can live out of water—some for hours, some for days, and some for years! Mudskippers hop around on land and even climb trees. So do climbing perch. Walking catfish can crawl along the ground and breathe air for a few days at a time. How do they do it? They have a special chamber in their gills to help them breathe.

Some fish have gills—and lungs, too! Its lungs help the lungfish spend a lot of time on land. In summer, the streams where it lives often dry up. So the lungfish curls up in a ball of mud at the bottom of a stream. It may sleep there for months—or even for years—until the rains come again. While it is sleeping, the lungfish breathes air through a little hole it has made in the mudball.

MUDSKIPPERS

How can fish breathe in water?

Fish can breathe in water because of the way their bodies are made. Fish need to breathe the gas called oxygen in order to live. Oxygen is in the air and in the water, too. Land animals have lungs, which can take oxygen from air but not from water. Fish don't have lungs. They have gills. Gills can take oxygen from water.

When a fish breathes, it takes water in through its mouth. The water flows through the gills, which take oxygen out of it. Then the water goes out of the fish's body through openings behind the gills.

Can fish live in a frozen pond?

If the pond is frozen solid from top to bottom, then fish *cannot* live there. Solid ice will not give fish the oxygen they need to keep alive. But usually when we talk about a frozen pond, we mean one with just a covering of ice. This sheet of ice has water below it, where fish can live. They usually stay near the bottom of the pond, where the temperature is warmer.

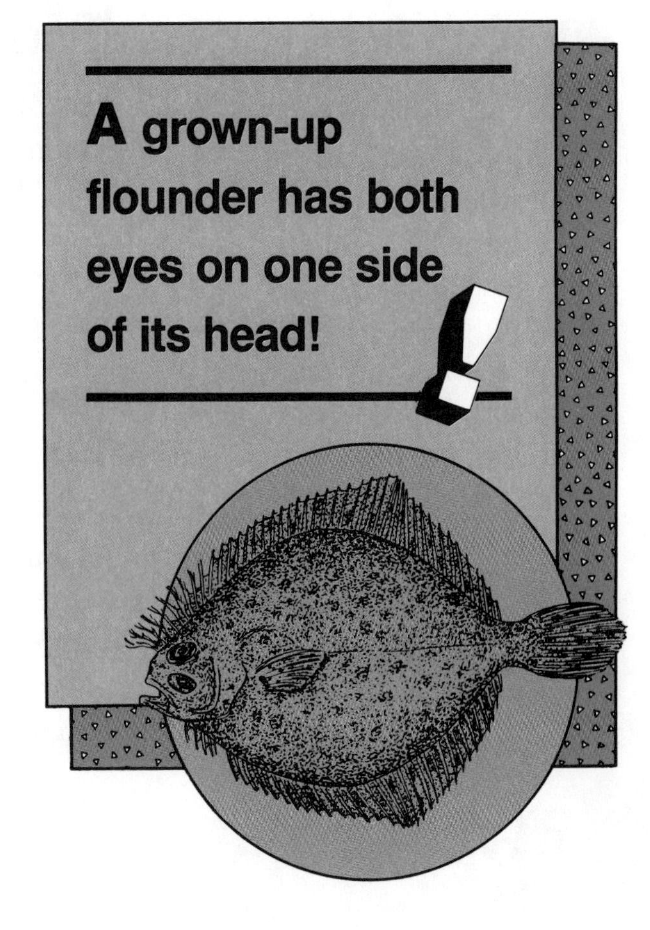

Do fish sleep?

Most fish do sleep—but with their eyes open! Fish cannot close their eyes, because they have no eyelids. When sleeping, many fish lie on their side or belly at the bottom of the pond, river, ocean, or aquarium where they live. The fish that don't sleep take rests. They just stop swimming and stay in one place for a while.

What do fish eat?

Because so many other water creatures are looking for food, too, most fish eat just about anything they can get. They eat insects, worms, and water animals, including other fish. Some even eat their own babies. There are fish that eat plants, too, but not many eat just plants.

Can fish make sounds?

A few can. A fish called the croaker makes a deep, grumpy-sounding *gur-rumph*. The sound is made in the fish's belly and is a lot like the noise a bullfrog makes. The grunting catfish makes a sound, too—but only when you take it out of water.

Does a fish feel pain when caught on a hook?

A hooked fish probably feels very little pain. In order for any animal to feel pain, it must have many nerves in the area that is hurt. The nerves send a message of pain to the animal's brain. A fish has very few nerves in its mouth, where it usually gets hooked. So it cannot feel very much there.

What fish is the smallest?

The pygmy goby is the smallest adult fish. It hardly ever grows longer than one-third of an inch, which is only this long: ___.

Which fish is the biggest?

The whale shark is the biggest fish. It can grow up to 59 feet long, and it can weigh up to 15 tons—more than twice as much as an African elephant! Fortunately for the other fish, the whale shark eats only plankton (PLANK-tun). Plankton is made of tiny, tiny bits of living matter that float in the sea.

SUPER SWIMMERS

Do all fish swim?

One kind of fish does not swim. It walks along the sand at the bottom of the water. This is the batfish. The batfish lives in shallow saltwater. Its fins are not really fins—they are more like legs. The batfish uses them to walk around.

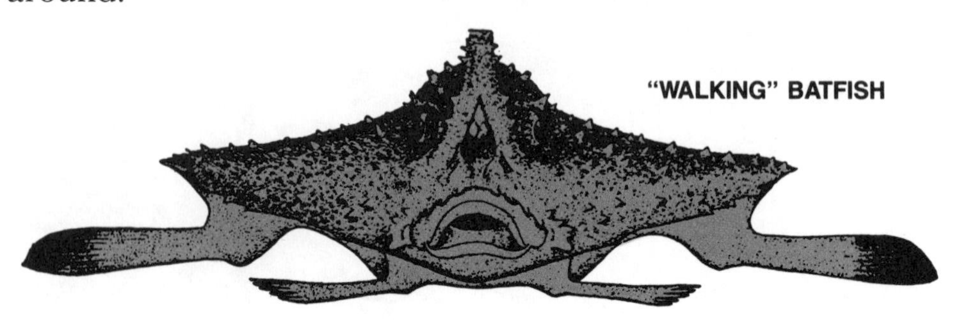

"WALKING" BATFISH

How fast can fish swim?

The fastest fish is the sailfish, which sometimes swims at more than 60 miles an hour. A few fish can swim between 30 and 45 miles an hour. Most are much slower. A small trout moves along at only 4 miles an hour, but it still swims faster than you do!

"SPEEDING" SAILFISH

What is a school of fish?

A school of fish is not a classroom. Fish schools are groups of fish that stay together. In a school, fish have more protection against hungry enemies. Each school is made up of only one kind of fish. You will never find bluefish and herring together in the same school. Nor will you ever even find baby fish in the same school as adult fish.

How many fish are in a school?

The number of fish in one school can vary. You might find twenty-five in a school of tuna—or hundreds of millions in a school of sardines.

SHARKS AND PIRANHAS

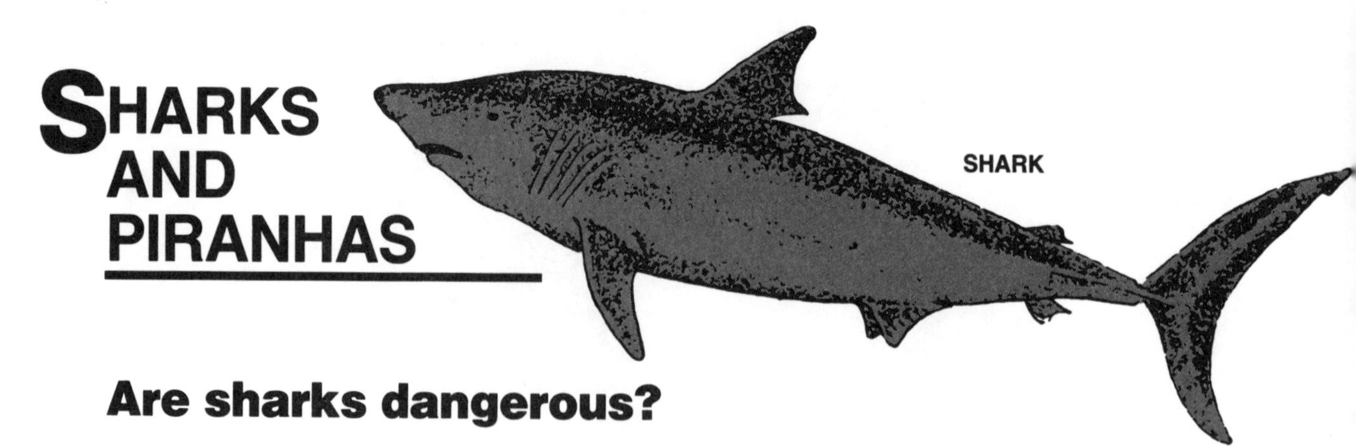

Are sharks dangerous?

Yes, many of them are dangerous, but some small ones aren't. The great white shark is especially dangerous. This shark has made many attacks on humans, mostly in Australia. It is strange to think that sharks have no real bones inside but often have terrible jaws and teeth. Even a shark's skin is dangerous. It is covered with tiny sharp spines that are like little teeth. You can get hurt just brushing against a shark.

Is any fish more dangerous than a shark?

Some people think that the piranha (pih-RAHN-yuh) might be more dangerous than a shark. Although piranhas are small, they have very sharp teeth. These fish travel in schools of thousands and attack all at once. In just a few minutes, a school of piranhas can eat all the flesh off a big fish or off an animal that falls into the water. Piranhas live in only one region—the Amazon of South America.

SEA SERPENTS AND ELECTRIC EELS

Are there any sea serpents?

Yes, there are, but they are not monsters. They are simply snakes that live in the sea or fish that have snakelike bodies. One of these fish is the oarfish. It grows to be 25 or 30 feet long and has bright red spines sticking out of its head. It looks pretty frightening, but is really quite harmless.

What do baby eels look like?

Baby eels don't look at all like their parents. They look like tiny glass leaves. As they grow, they change into the long, thin fish we recognize as eels.

Does an electric eel really make electricity?

Yes, it does. This fish's body is something like a car battery. It makes and stores electricity, which the eel can turn on and off. An electric eel uses the electricity in its body to catch food and to scare off enemies. The shock the eel gives can be strong enough to throw a man across a room. Small water animals are stunned by the shock and can't get away from the hungry eel. Scientists are still trying to find out exactly how this fish makes its electricity.

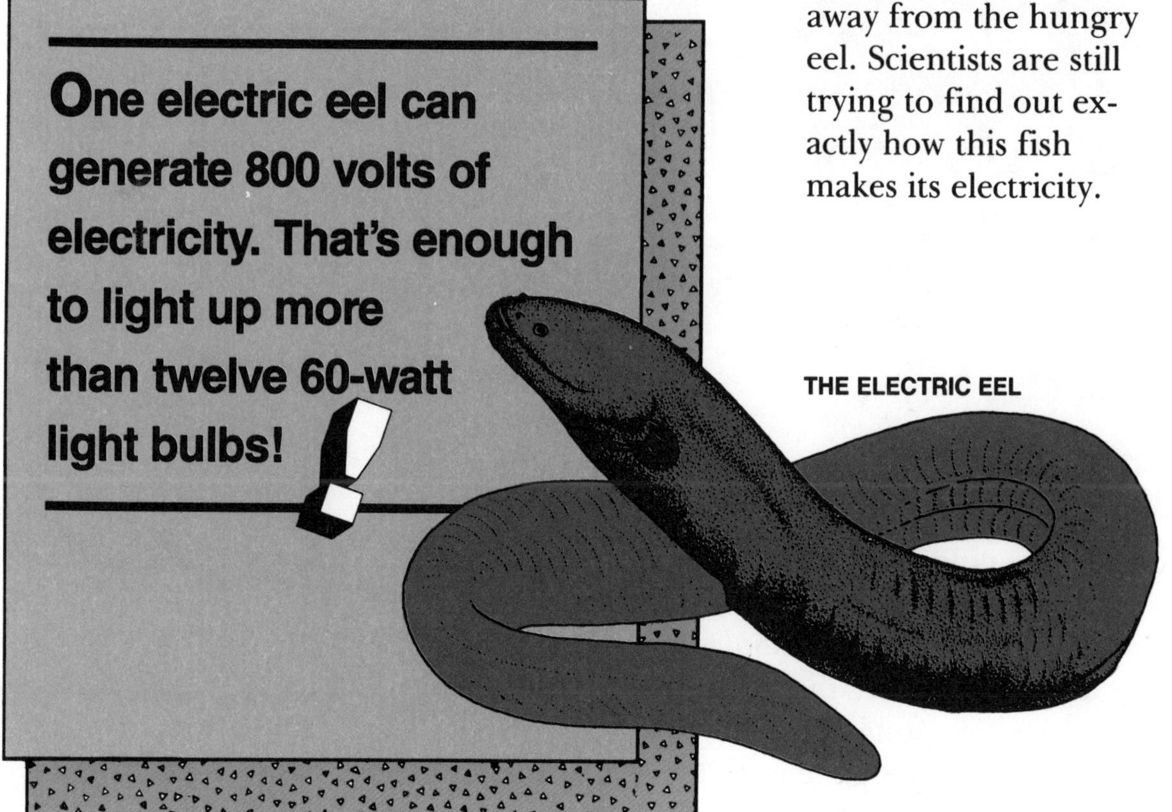

THE ELECTRIC EEL

FLYING FISH, GOLDFISH, SEA HORSES, AND MERMAIDS

FLYING FISH

GOLDFISH

Do flying fish really fly?

No, flying fish do not really fly. They glide through the air. Flying would mean that they flapped their fins the way a bird flaps its wings, but these fish don't move their fins when they are out of the water. They simply spread wide their large fins and sail through the air at great speed. Flying fish glide above the water in order to escape from their enemies, which are mostly dolphins.

When a flying fish wants to glide, it swims very quickly to the top of the water. As its head comes out of the water, the fish gives a powerful flip of its tail. This pushes the fish into the air. It can glide above the sea for up to 300 yards at a time.

In a large pond, a goldfish can grow to be as long as your arm!

How long can goldfish live?

At least one goldfish is known to have lived 40 years. Most goldfish can live about 17 years. Pet goldfish in aquariums don't usually live that long. They often die young from dirty water or a sudden change in water temperature.

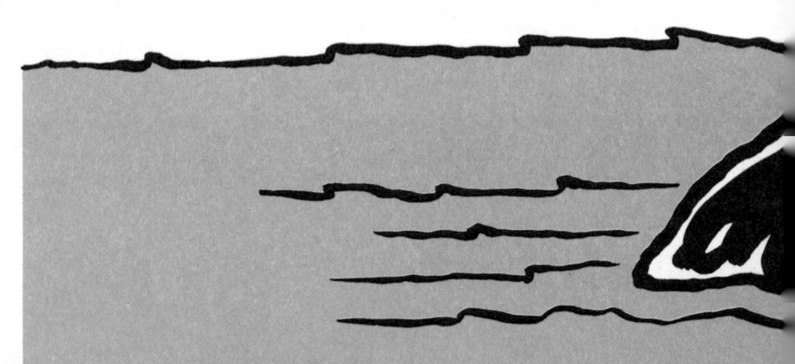

Is a sea horse a fish?

Yes, a sea horse is a fish, even though it doesn't look much like one. Except for its head, it doesn't look much like a horse, either. A sea horse doesn't move the way most fish do. It swims in an upright position, with its head up and its tail pointing down. The one fin on its back moves very quickly and pushes the sea horse along in the water.

Woodstock "rides" a sea horse.

What is a mermaid?

The word *mermaid* means sea maiden. Mermaids are supposed to be beautiful sea creatures who are half human and half fish. But they exist only in fairy tales. Legend says that Christopher Columbus thought he saw a mermaid while on a voyage in the Atlantic Ocean! What he probably saw was a manatee, or sea cow, a huge sea creature with two flippers and a spoon-shaped tail. Some people say its form somewhat resembles a mermaid's.

Baby sea horses hatch inside a pocket on their father's belly!

CHAPTER 17

Let's take a walk along the sand and surf and visit some other unusual creatures that live there. What makes some of these animals of the sea so special? Well, it's not what they have. It's what they don't have. They don't have bones!

NO BONES ABOUT IT

ANIMALS THAT LIVE IN SHELLS

What are seashells?

Seashells are the hard, protective cases that certain sea animals form around themselves. Oysters, mussels, clams, scallops, and snails are some animals that have shells.

Usually, the shells are empty by the time you spot them on the seashore. The animals have been eaten by other sea animals or by sea gulls. The shells are often pretty, and it's fun to collect them.

How do snails walk?

Snails, like most shellfish, have no legs. But most shellfish have a foot. The whole bottom part of a snail's body is one smooth, flat foot. The snail pushes its foot out from under its shell and pulls itself along a surface. As the snail moves, its foot gives off a liquid. This liquid helps the snail move more easily.

SNAIL

Can you really hear the sea in a seashell?

No, you can't. When you hold a large spiral-shaped shell to your ear, you hear a roar. But it's not the roar of the sea. The shape of the shell makes any slight sound in the air echo back and forth inside the shell. Sounds that you normally hear are picked up by the shell and made louder.

Some sea clams can move very quickly through the water by shooting a jet stream of water from their double shell!

SEA CLAM

Can a crab grow new parts if a piece of its body is cut off?

Yes. If you pick up a crab by one leg, it may let its leg drop off in order to escape. It then grows a new leg. Some crabs grow more than one new leg, which explains why fishermen often catch crabs with extra legs.

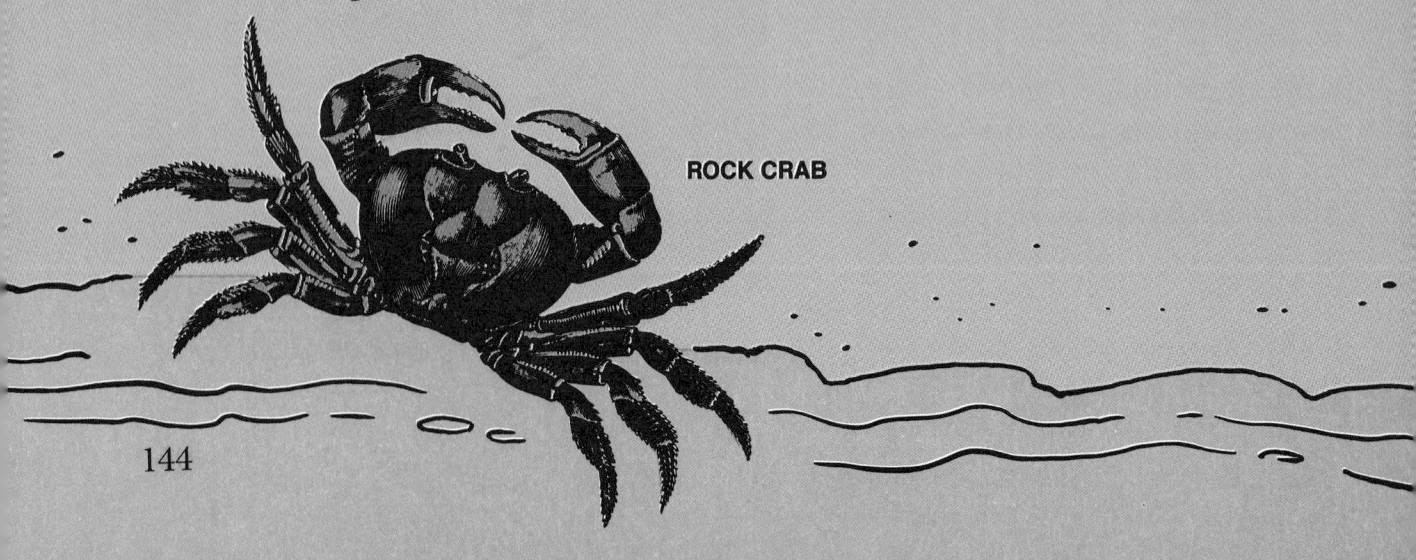

ROCK CRAB

How does an oyster make a pearl?

Sometimes a little grain of sand gets inside the shell of a pearl oyster. The sand rubs against the soft body of the oyster. To stop the rubbing, the oyster wraps the sand in layer after layer of the same shiny coating it makes to line its shell. We call this coating "mother-of-pearl." Gradually the bit of sand is wrapped in so many layers that a little ball forms. This ball is a pearl. Today, many of the pearls that are sold are produced by raising oysters and deliberately putting a bead inside the shell. This causes the oyster to make a pearl around it. We call these pearls "cultured."

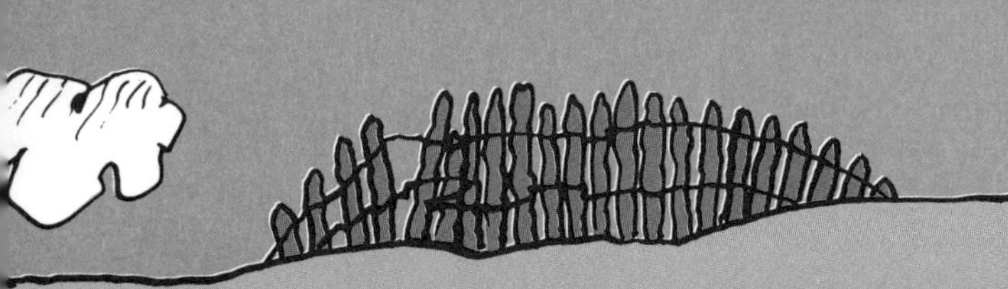

What is a horseshoe crab?

The horseshoe crab isn't a crab at all. It's much more closely related to a spider. With a shiny shell as big as a man's helmet and lots of skinny legs underneath, the horseshoe crab looks fierce. But it's really quite harmless. Horseshoe crabs have been on the Earth for millions of years.

What is a hermit crab?

A hermit crab is born without a hard shell. Because it doesn't have a shell of its own, it has to find a shell from another animal that will fit. The shell becomes the crab's home and protects it from its enemies. When it outgrows the shell, the crab has to find a new home, a larger shell from another animal.

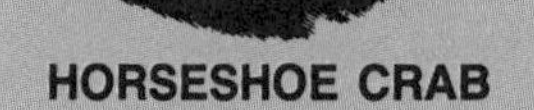

HORSESHOE CRAB

What are barnacles?

Barnacles are a kind of shellfish that spend their adult lives fastened to one spot. Some attach themselves to rocks. Others cling to crabs, whales, sharks, and ships. One ship can be the home for more than 100 tons of barnacles! Sometimes barnacles have to be scraped off ship bottoms because their weight and roughness slow the ship's speed.

THE AMAZING OCTOPUS

Is an octopus a fish?

No. An octopus lives in the sea, but it is more closely related to clams and other shellfish than it is to real fish.

Why does an octopus squirt black ink into the water?

An octopus squirts black ink in order to cloud the water. Then the octopus can hide from an enemy. That enemy may be a shark, a whale, or a person.

What does an octopus do with its eight arms?

An octopus uses its eight arms to catch crabs, clams, lobsters, and other shellfish. It also uses its arms to break open the shells of shellfish so it can eat them. On the underside of each arm are round muscles that act like suction cups. These can hold on to anything the octopus catches.

OCTOPUS

MORE BONELESS WONDERS

Is a sponge really an animal?

Yes. It is an ancient animal that lives in the sea. This animal has no legs, arms, fins, or stomach, and it doesn't move around at all. But each sponge has lots of tiny holes called pores. As water passes through these holes, the sponge takes out food particles. That's why sponges are sometimes called the filter of the sea.

A sponge can hold a lot of water, so people have long used sponges for cleaning. Real, natural sponges are still used for some kinds of work, but today, the sponge you use to wipe up a spill was probably made in a factory.

STARFISH

What is a starfish?

A starfish is a sea creature shaped like a star. Most starfish have five arms. They come in incredible colors—red, blue, even hot pink! If you cut a starfish into pieces, each piece will grow into a new starfish.

The basket starfish has more than 80,000 arms!

What sea animal looks like a flower?

SEA ANEMONE

The sea anemone (uh-NEM-uh-nee) looks like a flower, or at least like some sort of plant. It is a very simple animal—just a hollow tube with a mouth at one end and a lot of wavy "arms" around the mouth. The arms are used to capture food. Sea anemones come in a variety of colors—red, green, brown, and orange. Some have dots and some have stripes.

Can you eat a sea cucumber?

Yes, but it doesn't taste like the cucumber grown in a garden. In fact, a sea cucumber isn't a vegetable at all. It's an animal that lives at the bottom of the sea. We call it a sea cucumber because it is long and thin and looks like a cucumber. Although you won't find these cucumbers in a salad, you can eat them at some Chinese restaurants.

What is a jellyfish?

A jellyfish is a sea animal with a soft body and no shell. Some jellyfish are made up of several animals with different shapes that swim together and act like one creature. The Portuguese man-of-war is such a creature. If you meet a jellyfish in the ocean, you'll recognize it. It looks just like jelly!

Why do jellyfish sting?

Jellyfish sting in order to get food. First the jellyfish paralyzes a small animal with its sting. Because the animal then cannot move, the jellyfish can grab it and eat it. When you are swimming in the ocean, you may bump into a jellyfish and get stung. The sting may hurt, but you won't be paralyzed. So don't worry—the jellyfish will never eat *you*!

CHAPTER 18

Are they super spies or secret agents? Not these animals. But they live on land *and* in the water, so we say they have double lives. They are called amphibians.

ANIMALS WITH DOUBLE LIVES

FANTASTIC AMPHIBIANS

What are amphibians?

Amphibians (am-FIB-ee-uns) are animals that are born with gills for breathing in water, just like fish. Later, most of them develop lungs for breathing air. Most live in the water when they are young. After they have grown up they live on land, although they return to the water to mate and lay eggs. Like fish, amphibians are cold-blooded. When an animal is cold-blooded, its body has the same temperature as the air or water around it.

The amphibians include frogs, toads, salamanders, and caecilians (see-SIL-ee-unz). Caecilians are blind, wormlike animals that live underground when grown.

FROGS

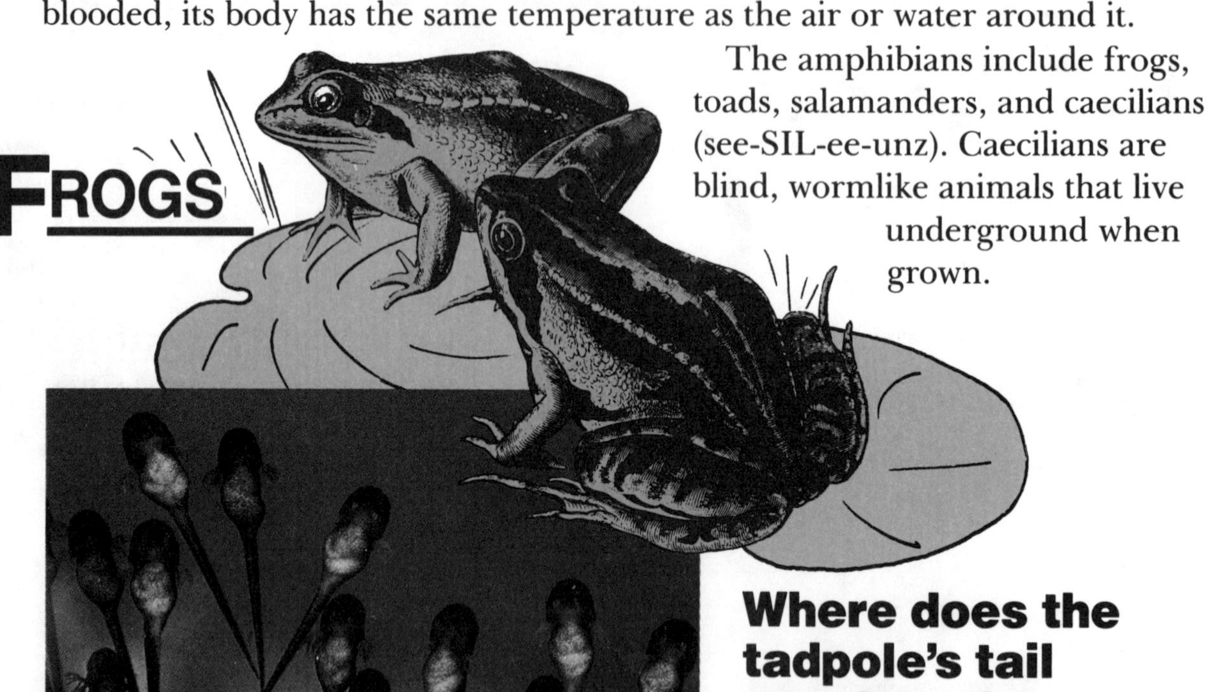

Where does the tadpole's tail go when the tadpole becomes a frog?

As a tadpole changes into a frog, its tail seems to get smaller and smaller. But the tail is not really shrinking. It is changing. It is slowly becoming part of the rest of the tadpole's body. During this time of change, the tadpole grows legs. Its gills change into lungs so it can breathe air.

What is a tadpole?

A tadpole is a baby frog or a baby toad, but it looks more like a fish. Tadpoles have no legs, and they have long tails. They breathe through gills the way fish do.

What do frogs eat?

Luckily for us, frogs eat mosquitoes. They also eat flies, moths, beetles, small crayfish, and worms. A frog's mouth is very large. It has two rows of teeth on the upper jaw and none on the lower jaw. A frog has a long sticky tongue attached to the front, not the back, of its mouth. This tongue can be flicked out quickly to catch insects.

What is the world's largest frog?

The largest frog is the Goliath frog of West Africa. The biggest one ever caught weighed more than seven pounds and was more than 32 inches long with its legs spread out behind it.

Do people really eat frogs' legs?

Yes, many people enjoy eating frogs' legs. The large hind legs—the jumping legs—are the ones eaten. They are usually cooked in butter. Many restaurants have frogs' legs on the menu. Frogs are even raised on frog farms to supply the demand for this unusual dish.

How far can a frog jump?

The longest frog jump on record is 17 feet and 4 inches.

TOADS

What's the difference between a frog and a toad?

TOAD

A toad is usually a chubby creature with rough, bumpy skin and no teeth. A frog is thinner, has smooth skin, and usually has teeth. Like all amphibians, frogs and toads are born in the water and return there to mate. Many frogs, however, also spend a large part of their adult lives in the water, while most toads do not. A frog's eggs are often found in big clumps in the water. A toad's eggs are often found in long strings, like beads.

FROG

Can you get warts from a toad?

No, you cannot get warts by touching a toad. That is just superstition. The rough skin of a toad looks as if it is covered with warts, and that is probably why the story got started.

However, the toad is not completely harmless. When a toad is attacked, it sends out a liquid poison from the bumps on its skin. The poison hurts the attacker's mouth and may keep it from eating the toad. If you catch a toad, and it lets out some of this liquid, be careful not to rub your eyes. The liquid will make them sore.

SALAMANDERS

What is the biggest amphibian?

The world's largest amphibian is the giant salamander of China and Japan. One found in 1920 was 5 feet long and weighed almost 100 pounds.

Where do salamanders live?

Adult salamanders are never far from water. They die if they can't keep their skin moist. Some grown salamanders live in ponds and streams. Others live on land, in damp places that are cool and dark. You can find them in shady woods. Often they lie under stones or in hollow logs.

What are mud puppies, newts, and efts?

Mud puppies and newts are simply kinds of salamanders. During the time newts are living on land, they are called efts. As efts, they are orange colored. When they go back to water to mate, they turn green.

SALAMANDER

DID YOU KNOW...?

• Camouflage (KA-muh-flodge), blending in with your surroundings, can be very important in the animal world. The inchworm is a real camouflage artist. When an enemy such as a bird comes near, the inchworm freezes and pretends to be a twig. It blends in so well that the bird will often walk right across the "twig" and miss its lunch.

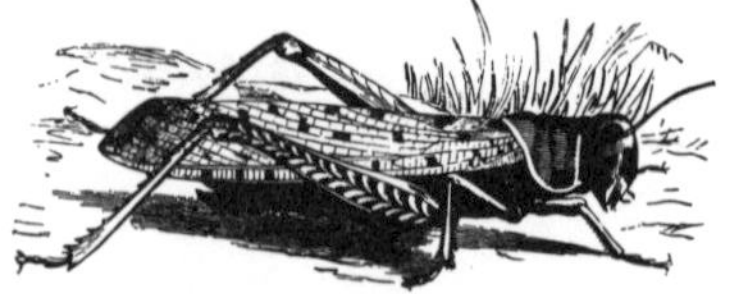

• Crickets don't use their mouths or throats to make their chirping sound. They rub their wings together. Only male crickets make this sound. They attract female crickets with it.

• The amazing earthworm doesn't have to worry when it gets cut in two. The pieces will wiggle around for a while, then the smaller piece usually dies. The larger piece often grows back the sections it lost—if it lost just a few. This is possible because an earthworm's body is made up of a long row of sections that are all pretty much the same. When a few sections are cut away, the worm can replace them.

• The longest earthworm ever found was nine feet long! It was found in Australia.

• The leafy seadragon looks more like a plant than a fish. Its body and fins look like branches of seaweed. It lives in the ocean off the coast of Australia.

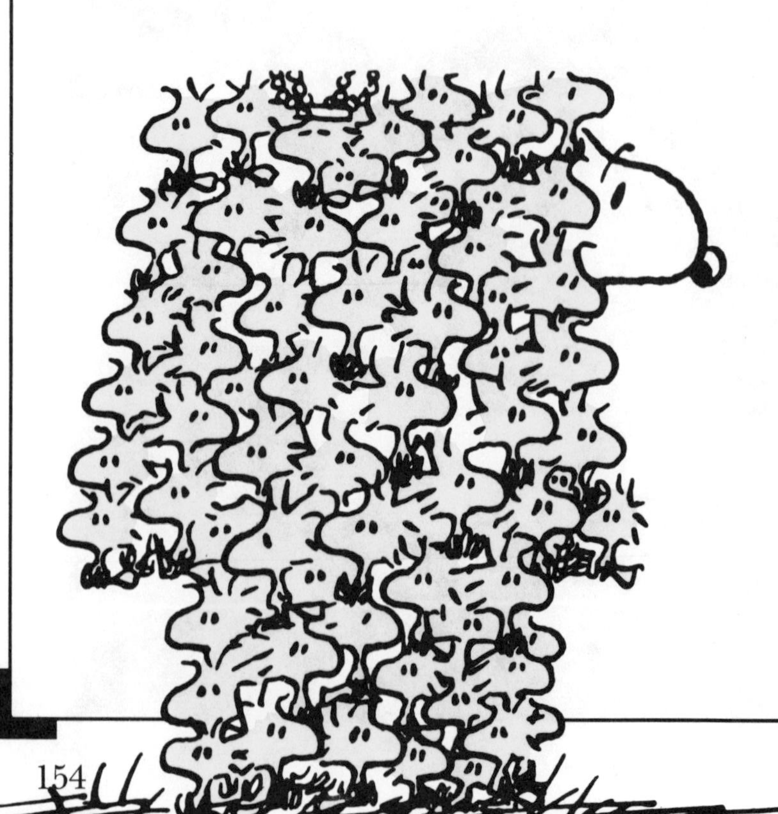

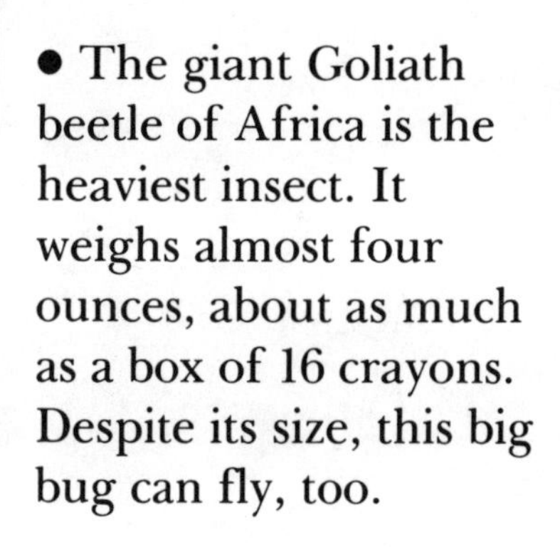

• The giant Goliath beetle of Africa is the heaviest insect. It weighs almost four ounces, about as much as a box of 16 crayons. Despite its size, this big bug can fly, too.

• The dolphin is one of the world's fastest swimmers. Some dolphins can swim 40 miles an hour! Playful and friendly, the dolphin loves to swim alongside ships. And this great swimmer isn't even a fish. The dolphin is a mammal.

• Most animals have to spend a lot of time finding food and shelter, but parasites depend on other animals. Some dogs are experts on one of the best-known parasites—fleas!

• When honeybees dance on the face of the honeycomb, it's not just for fun. They are really letting other bees from their hive know where to find food. Fellow workers watch the honeybee dance and are able to get directions that will lead them to the exact location of the pollen and nectar.

• Many animals know how to do some things from the minute they're born. Fish are born knowing how to swim, and no one has to teach a spider how to build a web. This special knowledge an animal is born with is called instinct. Don't you wish you were born knowing how to do math problems?

• Some fish that live deep in the ocean have their own sources of light. These fish have special parts under their eyes that act like little light bulbs helping the fish see their way through the pitch-dark ocean. What are they called? Lantern-eye fish!

• Puffer fish look quite ordinary when they are swimming along. But when they are threatened by an enemy, they can expand to several times their normal size. Filled with water, the puffer fish look much bigger, and the enemy is frightened away.

There's much more to discover in Snoopy's World.
If you've enjoyed *Creatures, Large and Small*,
you'll want to read...

How Things Work

People and Customs of the World

Earth, Water and Air

Land and Space